❝Whether you tweet, text, or yammer, *Read Between the Lines* will fast-forward your ability to communicate in the new world. I thought I was an early pioneer in texting. By reading *RBTL* I learned more in thirty minutes than I had in five years of finger typing. More importantly, I learned how to look cool to my children. You need to get to the point, be clever, and figure out how to take advantage of the limited space (characters) provided by services like Twitter. The new world is here **NOW**, and you better get hip or get deleted. **❞**

Peter H. Jackson
Silicon Valley CEO & Director

❝My doctoral research shows that there is much more to texting than abbreviations and phonetic spelling: Texting is a creative, expressive, and practical outlet for a surprising range of uses. In short, texting deserves much better press than it usually gets! It's great to see a book like *Read Between the Lines* address texting in such an entertaining and informative way. **❞**

Dr. Caroline Tagg, PhD
First person to receive a PhD in texting,
Birmingham University, U. K.

"We believe that public safety must be prepared for and utilize technology that is available today and in the future. Cell phones and text messaging have become a mainstay in our society; thus public safety must adapt to this new avenue of communication. Because we must reach out to all of our citizens, text messaging opens many doors. We are very proud of our small role in becoming the first in the nation to provide this service to the citizens of Black Hawk County, Iowa.**"**

Thomas Jennings
Chief of Police, Waterloo, Iowa
First U. S. county to implement 911 texting services

"I like the flexibility texting offers. I can ask a question as I go along in a class and get students to respond quickly and en masse. All I need to have is the number they text to, and I don't need pre-prepared Power Point slides with the question and a choice of answers.**"**

Dr. Kevin Linch, PhD
Pioneer of lecture texting within a university setting,
Teaching Fellow, School of History, University of Leeds

Read
Between the
Lines

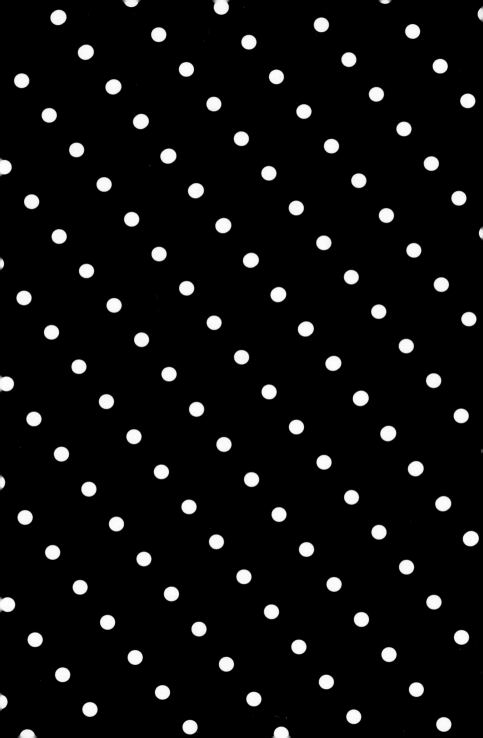

Read
Between the
Lines

A Humorous Guide to Texting with Simplicity and Style

Shawn Marie Edgington

Read Between the Lines

A Humorous Guide to Texting with Simplicity and Style

© 2010 Shawn Marie Edgington

Manufactured in the United States of America.

For information, please contact:
Brown Books Publishing Group
16200 North Dallas Parkway, Suite 170
Dallas, Texas 75248
www.brownbooks.com
972-381-0009

A New Era in Publishing™

ISBN-13: 978-1-934812-53-2
ISBN-10: 1-934812-53-6

LCCN: 2009936684
1 2 3 4 5 6 7 8 9 10

The Dedication

Many thanks to all of you who helped in the creation of **RBTL**, including all the teens, parents, and colleagues who shared their texting tricks, trials, and tribulations over the years.

To Margarete Gockel, your incredibly brilliant illustrations just make me want to stand up and "do a little dance" every time I see them! And a great big thanks to my entire team at Brown Books. You guys and gals are extraordinary!

Special thanks to my husband David, and to my incredibly amazing teens Nicole and Derek for all of your loving support, and to the other two "golden" boys in my life, Dakota and in memory of Brinkley, my loyal companions. To my BFFs Wendy, Pam, and Cheryl for being so fabulous in every way!

Finally, my warmest thanks to my mom, Patricia Dolan, who always gives me the encouragement I need to keep moving forward no matter what, and who constantly inspires me to dream big and always believe.

~Shawn Marie Edgington

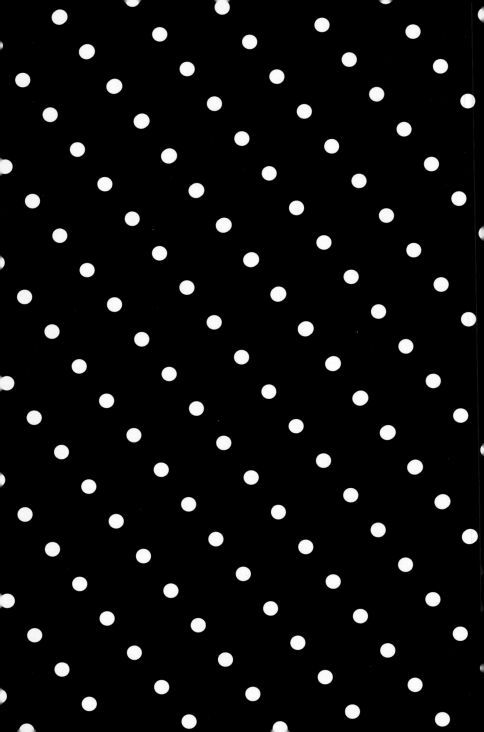

The Goods

The Author's Note. xi

The Quick Start Guide. xiii

  1. Post-Its on Steroids. 1

  2. Fast and Furious Teens 11

  3. Parental 411. 21

  4. From the Bedroom to the Boardroom 37

  5. Just Text It: Go Green, Text Smart. 49

  6. Text Benefits for Anyone and Everyone 59

  7. Corporate Texting–

        From Main Street to Wall Street 75

  8. Texting Tips and Tricks for the Text Savvy . . 87

  9. Text Your Tweets to Stardom 95

  10. The Golden Rules of Texting 117

The List. 129

   The Top 50 130

   Text Lingo Dictionary 133

Mind-Blowing Texting Info and Fun Facts 161

The Author 165

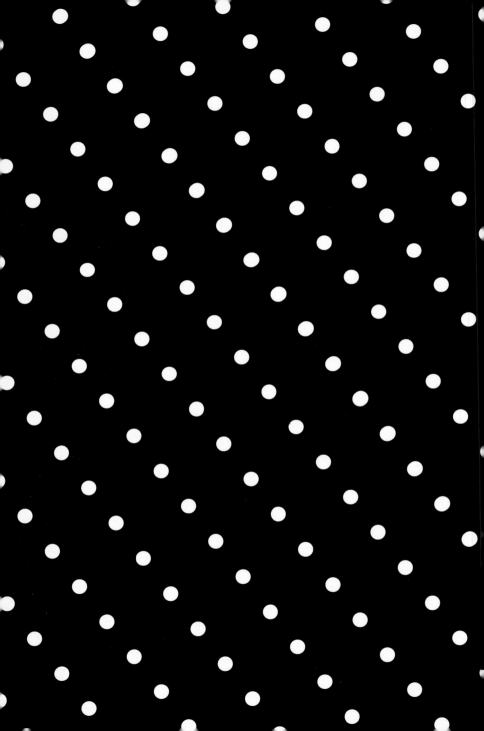

The Author's Note

Read Between the Lines, a fun and not-too-technical text survival guide, allows you to master the explosive language of instant, interactive communication that text messaging offers. With this hip and humorous, purposeful and portable, quick reference guide, you'll be able to keep family ties firm, maintain strong friendships, understand why text manners matter, uncover "need to know" parental information, learn texting tips and tricks, and discover other fun and useful information that will guide you towards text message success.

Read Between the Lines takes the mystery out of abbreviated words by providing you with a text message road map. You'll also find important dos and don'ts, situational text etiquette, how to text your tweets, and a comprehensive dictionary of popular text lingo. This text survival guide is designed to show you how to text and how to avoid the common mistakes and challenges that often arise from being a NOOB (newbie) in the text world.

Twenty-first century technology allows us to stay in constant contact with anyone, anytime, anywhere. Within a few short years, there will be trillions of text messages sent and delivered globally every day, so now is the time to join the texting phenomenon with style, finesse, and a bit of sarcastic grace with Read Between the Lines as your "WHEN, WHERE, and HOW TO" text guide.

The Quick Start Guide

Take a not-too-serious look into the world of text messaging. Soon, you'll quickly understand the advantages text messaging provides. After all, texting is topping technology's heat index as the most widely used data application on the planet, with over three billion active users worldwide.

Do you think text messaging is just for the teen generation or too complicated for you to learn? In reality, it's an incredibly easy and accessible communication tool that rests in the palm of your hand. A text message gives you an instant connection when a phone call might not go through, because a text requires a lower service and battery level than a call.

With no expertise required, text messaging makes bridging the communication gap with your friends, family, and colleagues much easier, not to mention more effective and efficient!

Let's Get Started!

Before you start to travel down the path of your new communication highway, grab your cell phone and see how easily you can send a text following these four simple steps:

1. Choose to compose a Text or SMS Message from your cell phone's main menu. Note: SMS = Short Message Service. Almost all phones have SMS capability built in.
2. Type your message (up to 160 characters per message) using the keypad on your cell phone.
3. Enter the ten-digit cell phone number of the recipient. (Area code + number)
4. Hit the "send" button.

If you already have your buddies' numbers in your address book, texting is even easier:

1. Select the person you want to call, and instead of hitting "send" to dial the number and make a voice call, select Text or SMS to the ten-digit cell phone number.
2. Type in your message (up to 160 characters per message).
3. Hit the "send" button.

That's it. Really—it's that easy! :)

OMG–U Can Txt!

U Rock! You sent your first text! Great work. Now it's time to respond to your first text, preferably with style.

Here are a few fun examples to make a statement and send a memorable response to your first text message, or flip forward to the section with the most popular text lingo and pick what works best for you.

C u ltr @ home 2nite! :)
See you later at home tonight!

OMG, Im soooo proud of u-H&K
Oh my God, I'm so proud of you, hugs and kisses

U can do it! U R amzing!
You can do it! You are amazing!

LOL...TTYL :)
Laugh out loud, talk to you later

HAND
Have a nice day

Plz Dnt b L8t 4 Dnr
Please don't be late for dinner

Even if you don't plan to use text lingo as your SOP (standard operating procedure), it's a great sense of accomplishment not only to learn a new technology but also to know how to use it as a member of the Text Generation would.

Talk's Not Cheap

The best things in life are almost free, or darn cheap! With texting, talking is no longer cheap, especially when you compare the cost of texting to calling. Cell phone plans can be reasonable, but minute overages can be brutal. Revamp your cell phone bill with a plan that allows you to utilize text messaging as an affordable alternative to burning up the airwaves.

Pick Your Plan

WARNING: Sending and receiving texts can get very costly if you don't have a text messaging plan set up with your wireless provider.

The sky's the limit when it comes to the number of text plans offered by the various providers. For example, you can purchase packages that include 100, 200, 500, 1,000, or unlimited monthly text messages. There are single user text plans, family text plans, plans that include data and text, and plans that include just text messaging.

Start with deciding how many texts you plan to send/receive each month multiplied by the number of users who will be texting. Compare costs for what's offered with your provider, and then pick your plan. Whatever you do, don't underestimate text usage, especially with a tween/teen in the house.

If you plan to be the next textonista, you'll want to consider an unlimited text messaging plan for yourself. Most cell phone service providers offer unlimited text message packages for a monthly flat fee that will let you send and receive text messages 24/7; you'll never have to worry about going over the text limits within the U. S. boundaries.

To Infinity and Beyond

Going global? For international long-distance texting plans, you'll need to call your provider prior to your trip to add the option that's specific to your travel plans, including the country you are destined for or want to exchange text messages with. International text messaging will cost more than services provided in the U. S., but it still offers a great alternative to the cost of international phone calls.

International long-distance text message plans are a good idea to consider for those who text from the U. S. to other countries. ILDTM allows you to send text messages to international numbers in over 140 countries at a significant savings over not having a plan.

No-Cost Texting

Most major wireless providers allow anyone to send and receive text messages through their corporate Web site at no cost to the sender. It's a nice, no-cost option, but it won't allow you to send/receive messages from your cell phone, which is one of the key reasons texting is such a great tool.

You'll also find a wide range of Web sites that offer free text messaging services via the Internet. Basic services allow you to send texts free of charge, while other services (with more bells and whistles) allow you to set up a free account so you can send and receive free texts on their site. This feature is especially useful for those who don't have a cell phone, like the hard of hearing, who still want to send and receive texts from friends and family who have cell phones.

Since we all have different needs, consult with your wireless provider and select the plan that best fits your specific requirements. Ask questions, make notes, and check pricing with at least one other provider. You could be surprised at the difference in costs between the two, especially when adding additional options to your plan.

911–In the Text of Time

Text messages almost always get delivered, which is particularly important during an emergency. During a catastrophic emergency like an earthquake, phone lines can be jammed. Everyone's trying to get in touch with loved ones to see if they're okay or if they need help.

Only one or two bars? Text messages don't need two bars to be sent. Instead, they stand in queue and are delivered in the order they're received. During a catastrophic emergency, your text may take longer to get through, but you can be virtually assured that it'll be delivered quicker than a call.

Most 911 calls made from a cell phone aren't routed to your local emergency services. Instead, they're routed to a central calling location, where you might get put on hold, or you might hear "your call will be answered in the order it was received." This isn't the case if you send an emergency note through text messaging. Your texts will be sent directly to your city's emergency system, as long as your city offers the service. Contact your local city's emergency service division to see if it offers the ability to report emergencies by text.

Congratulations

You're ready for prime time! Once you begin to make texting one of your go-to methods to reach out and touch someone, you'll start to remember examples of word abbreviations and text lingo that will help you feel more comfortable texting. And whatever you do, don't worry about all the acronyms that come along with texting.

Think of texting as just another crazy-cool tool you're adding to your communication bag of tricks that does much more good than harm. You'll soon learn to depend on the many advantages texting offers—all available for your personal use, under your total control, always resting in the palm of your hand.

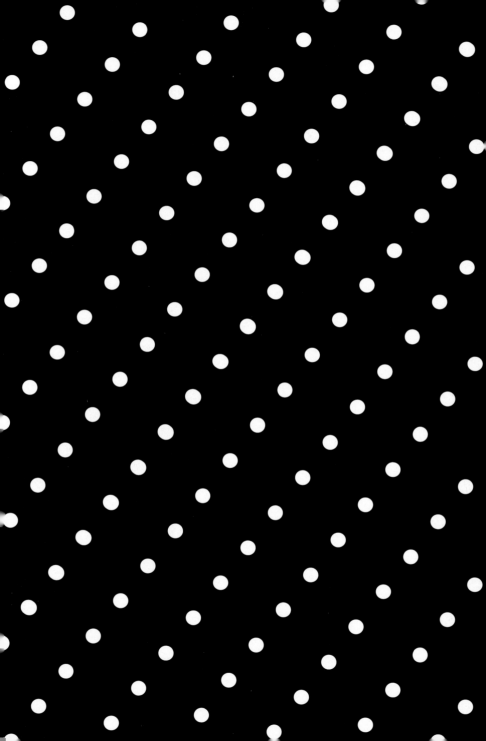

Dont frgt your hmwk . . .
c u 2nite! :)

– One –

Post-Its on Steroids

You're on a business trip and you're worried about how things are going at home without you. You check your watch and realize your kids will be leaving for school soon, so you text this message to your fourteen-year-old daughter (who is notorious for forgetting her homework). She texts back, **Thx Mom! U rock!!!** You smile and slip your phone into your purse. Texting helped you send your daughter a reminder from thousands of miles away. This is one morning when a Post-it just wouldn't do the trick!

Think of a text message as the new "shorthand" or an informal note you scribble on a piece of paper like a Post-it Note. It's a short message equivalent to about two sentences that you want to communicate informally. On the other hand, if you want to communicate formally using correct spelling and grammar, a text message is **NOT** the right choice. Opt to send an e-mail or even a note by snail mail.

Bridging the Communication Gap

Text messaging has taken our society's youth by storm, a trend that's quickly migrating across the generations. Puberty is no longer a requirement to text, so say **YES** to the undeniable force pulling you towards the text message phenomenon.

After all, you bought this book didn't you? What's the payoff? Texting will provide many valuable advantages over traditional lines of communication, as long as common sense is your common denominator.

Still not convinced? On the next pages are some texting benefits and, to be fair, a few of the drawbacks of learning to text. As you'll see, the positives far exceed the negatives.

The Upside of Texting:

- ⊘ **Saves Time**—it's quick
- ⊘ **Saves Minutes and Money**
- ⊘ **Super Simple**—easier than making a call
- ⊘ **Discreet and Private**
- ⊘ **A Quieter Interruption**—less intrusive than a call
- ⊘ **Instant Communication**
- ⊘ **Global Reach**—a cost-effective means to global communication
- ⊘ **Emergency Situations**—communicate when a call isn't possible

The Downside of Texting:

- ⊘ **Forgetting How to Spell**
- ⊘ **Text Addiction**—requiring a one-way ticket to t.a. (texters anonymous)
- ⊘ **Injuries, Including Thumb Numbness, Tingling, and Eye Strain**

Texting 101

Because a single text message is limited to 160 total characters, a universal shorthand has been developed to allow you to say more in a single message. Words become letters, abbreviations, or numbers, and letters are combined to sound out a single word.

What's the result? When texters put together letters and eliminate vowels, they create text slang, jargon, acronyms, and lingo, and the number grows hourly. Here are a few tricks to help you get started:

Don't Worry About Upper- and Lowercase Letters:

When you're texting, uniformity doesn't matter. Your phone will automatically default to upper- or lowercase, usually depending on the punctuation you use.

> Ill b at brooks til 7 will u pic me up?

Eliminate Punctuation:

Don't worry about those apostrophes or commas. Eliminate them where you can to shorten the amount of characters you use in your text. Keep your "?'s"—you'll need those.

> Lts go 2 the movies 2nite! dnr @ 5 ok or is 6 btr?

Eliminate Vowels:

A simple way to shorten words is to eliminate vowels. Text = TXT, Please = PLZ

Replace Words with Numbers:

For example, the word "For" becomes "4," or the word "Ate" becomes "8." You can also replace part of a word with a number. "Later" becomes "L8R."

Use Sounds and Letters to Represent Words:

For example, instead of writing the word "See," you would use the letter "C." For "You," you would use the letter "U," and "Why" becomes "Y."

Use Common Abbreviations for Common Words:

There are many common abbreviations most people know, so go ahead and use them. Here are a few samples: "Strbks" for "Starbucks," "thx" for "thanks," and "BTW" for "by the way."

Add Some Text Lingo:

At your discretion, you may add text lingo to your message out of the dictionary in the back of this book.

Combine All of the Previous Methods:

You can use one or all of the methods above in your text messages. As you become more experienced, use what works best for you.

Don't feel that you always need to use abbreviated language in order to text. It's really just a way to shorten the size of the words so you can say more within the 160-character limit for each message. Don't worry if you run out of room—your phone will allow you to continue your message beyond the 160-character limit by automatically creating additional messages for you.

Since many cell phones now have predictive text, preloaded text messages, and easy-to-use keypads, you can often spell out your entire message if you have enough room. Whatever you choose, you'll have fun creating your own personal savvy as a new texpert.

If you're a "Level One" texter and have children at home, grab your favorite tween or teen and ask for some helpful hints. You won't believe how quickly your image will improve in your teens' eyes when you surprise them with even your elementary texting style.

The Text Debate

There are many controversies about text messaging, text abbreviations, and text lingo, arguing whether they're ruining our youths' communication skills. Remember telegrams? The length of the message dictated the cost, so abbreviations were used without "ruining" the English language. Texting is really just an updated and more personal version of the telegram.

With every new technology comes skepticism. Imagine all the fuss when the printing press, radio, telephone, television, and computer were invented. These items have all become serious necessities for almost everyone.

The reality is that old debates will continue and new debates will begin. Fortunately, now that many cell phones have full keyboards, better predictive text recognition, and inexpensive/ unlimited text messaging plans, most texting concerns have been resolved.

Woo Hoo #1

You'll soon become the next Text-Master by entering into the new frontier of the text generation—at warp speed. Discover numerous ways to assimilate text messaging into your everyday life, allowing you to adapt to this new digital superpower, while at the same time reaching triple platinum communication success!

Quick Text Lingo Lesson

BRB · Be Right Back

B4 · Before

BTWN · Between

G2G · Got To Go

GR8 · Great

IDK · I Don't Know

POV · Point Of View

SOL · Sh** Out of Luck

TTYL · Talk To You Later

TTFN · Ta Ta For Now

:) · Smile or Happy

:(· Frowny Face or Unhappy Face

Where R U? Practice is out 1 hour erly . . . can U pick me up soon?

─ Two ─

Fast and Furious Teens

Your ninth grader sends you this message just as you are about to make a presentation to a new client. Frantically, you text your husband, who's in a meeting: **Dereks out 1 hr erly frm practice . . . can U pick hm up?** He manages to text back, **Can do, ill b thr in 15 mins.** You reply to your son, **Dad will B thr in 15 mins.** Problem solved with text messaging. What your family experienced is the new wave of text communication that every tween and teen has mastered and that you can as well.

Who IS Generation Text?

Texters between the ages of twelve and twenty-four make up Generation Text; they're quick, smart, fast-moving, and hard to keep track of. In this digital era, keeping the lines of communication open with your teens can have a major impact on your relationship.

In case you haven't noticed, the Text Generation is completely entwined and engaged with their most coveted asset: their cell phone. Teens use their phones to achieve total connectedness morning, noon, and night with anyone who's willing to join their network. In most cases, you'll find texting is their number one priority!

Generation Text uses text communication as an important part of their daily life. They want to be heard and understood on their own level and in their own preferred way. Teens feel that if they don't have their phone in their hand or in their pocket, they're completely disconnected and out of their communication loop.

Here are just a few examples of how texting can help you communicate with your fast and furious teens:

- **Reminder Texts**
- **Need a Ride**
- **Instant Answers to Quick Questions**
- **Teen Tracking**—find out where they are, whom they're with

- **Good Day**—great news, great job, I'm proud of you
- **Bad Day**—encouragement needed
- **Minds Change, Plans Change**
- **Emergencies**
- **Troubled Teen Text**
- **Time for Dinner Text** (yes, I do mean from/to the other room)
- **They Want to Tell You Something in Private**
- **Cry for Help Becomes Easier**

TXT 2 Connect

Most teens keep their phones on vibrate or silent, which means they don't hear their cell phone ring when they receive a call. If you're relying on e-mail to be the window into your favorite teen's world, or if you're thinking that if you call them they'll pick up, you haven't been given the 411 on teen communication.

What will actually happen when you give them a ringy-dingy? You'll probably get voice mail. What about the e-mails you've been sending for the last six months to your son in college? They're probably still sitting in his in-box, or maybe they went into a spam folder or filter. Bottom line: expect your voice messages not to be heard and e-mails not to be read. Even more common is that after your tenth voice message, you'll hear, "This voice mailbox is currently full"—more proof that not only are voice messages not listened to, they're completely ignored.

Now is the time for a paradigm shift. How about a guaranteed instant response to your nagging parental questions—without the attitude? Go ahead, send your teen a quick text and ask a silly question or something you would like a quick answer to.

You should get a response within seconds. The unwritten code of conduct for teenagers is to answer every text message within seconds of receiving it. Only when they're in the shower, in the classroom, out cold dreaming big dreams, or pining for their phone while it's visiting parental jail, would they willingly violate this teen text code.

Lingo Usage

Teens and tweens are the most likely to use text lingo, an abbreviated language that's been designed and customized to meet the immediacy of how text messaging works. With Generation Text, plan to see lots of **LOLs, OMGs,** and **IDKs.** If you're up to speed on some of the basic lingo that teens use, it's acceptable for you to slide some lingo into your messages here and there, which lets them know you're willing to communicate with them on their level, using their language.

Typical Teen Texts

You'll find that most teens use texting for most of their day-to-day chats with their friends, like:

> Wut r u up 2? Im w/ nicole @ the
> mall . . . can u meet us?

If you pay close attention, you'll quickly notice teens carry on text conversations with three or more friends, all at the same time. How is this possible? Talent and lots of practice!

Watch Your Wallet

Unsuspecting parents have been sent into complete craziness after opening that first monthly phone bill. Surprise! Before parents have even selected a text plan, their teen has hit the road running within the magical world of texting. They're shocked to discover a $597 bill—all from the 5,486 text messages sent the two weeks before. How could they possibly send that many texts in two weeks? Easy, **VERY** easy! Remember, teens rely on text messaging as one of their primary socializing tools, so they quickly rack up huge amounts of texts every month.

You might want to implement serious damage control by adding an unlimited text option to your wireless plan, avoiding the potential for any overage charges.

Teen Stats

As of the first quarter of 2009, according to Nielsen Mobile, U. S. teens (ages 13 to 17) have the highest levels of text messaging:

- **Teens average 2,900 text messages a month.**
- **Teens average 231 phone calls a month.**

These latest teen statistics are loud and clear. Teens use their phones to text 1200 percent more than they use their phones to make calls. Of course, there are those teens who far exceed the average, sending **MANY** thousands of texts a month. One thing's for sure, a text is much more preferable than a call.

Show Your Cool Side

Discretion has its advantages when communicating with teenagers. With text messaging, questions and comments can be kept on the DL (down low). Tweens and teens are usually surrounded by friends, so when you send them a text, they're more likely to text you back (along with the other seven texts they're responding to at the same time). Even if your teen is in the company of her five BFFs, and she wants to tell you something important, she can easily get her message off to you:

> Mom, can u plz come pick me up
> from the party…I'm not comfortable with
> what's going on here.

Not only will your teen be able to send this text in confidence, she'll also be able to tell her friends a "little white lie" as to why she's getting picked up early. Without being accessible by text, you could very well miss out on an opportunity to pull your teen out of an awkward situation.

Woo Hoo #2

Teen communication can be challenging for parents, but it's one of our highest priorities. Now is the time to embrace TMing and take advantage of your teens' constant connectivity. Discover how the world of texting can help you build stronger and more successful relationships with your little darlings. Every good relationship requires some give and take. Maybe this is the time you give a little by learning how to text.

Quick Text Lingo Lesson

BFF · Best Friend Forever
BTW · By The Way
HAH · Big Laugh or Laughing
JK · Just Kidding
LMAO · Laughing My Ass Off
LOL · Laugh Out Loud
OMG · Oh My God
ROFL · Rolling On
Floor Laughing
*$s · Starbucks
YTB · You're The Best
:/ · Confused Face

Mom, Im goin 2 *$s w/Marina,
thn 2 the movies, k?

-Three-

Parental 411

*Y*our daughter sends you this text message on Friday afternoon because she knows you're at work, busy with clients. You excuse yourself for a quick second, pick up your phone, and read the message. You smile because you're really happy your daughter keeps you in her social loop. You send her a quick text to say, **thx for the update, have a gr8 time, b home by 11.** Your daughter texts back and says, **kk...c ya @ home!** All of this was done in less than one minute, without having to hear a phone ring, answer it, or bark out orders to your daughter so everyone can hear. Nice job.

In the world of parenting, we can't even count the numerous ways we worry about our kids! If you have just a few seconds and a cell phone, learning to navigate the world of texting is a great way to keep the lines of communication open with your busy teenagers.

Text to the Rescue

Teenage communication between the ages of thirteen and seventeen is a lot like trench warfare. Teens typically want more and more independence from their parents, and parents want to stay in control and remain a positive influence on the decisions they're making. Can you feel the pain? Here's a few ways you'll benefit from text messaging, to help reduce your virtual scrapes and bruises:

- **You'll get a response to your text message, even when they're with their friends.**
- **Teens are more likely to provide quick check-ins and updates.**
- **You can strengthen your parent/teen bonds with communication they prefer.**
- **Tone of voice and facial expressions become invisible.**
- **Texting allows for teens' independence, without your losing parental control.**

Knowledge is power, especially for parents. With the ability to text, you can stay connected without helicoptering over their every move. Text your teens to find out whom they're with, what they're doing, where they are, or why they're late. It's a great way to find out information you want to know. The good news is that you'll get an instant reply that gives you instant relief without Rolaids.

In the Loop

Kids are comforted to know they have a quick way to express their ups and downs, keeping you involved in their busy lives. If you've added texting to your communication arsenal, they can send you quick notes to let you know what's up, what great thing just happened to them, or how your little drama queen thinks it could be the end of the world as she knows it. Texting is a powerful communication tool that opens a new window into your teens' lives.

Text vs. Call

It's late at night. You're tired. You definitely don't want to answer the phone, let alone hear it ring. When your teenagers are out and about, moving from place to place, they can quickly and easily text you in confidence, explaining where they've landed and what's next on their agenda. Because they're already sending out tons of texts (they never stop) sending a few extra texts to their parents or guardians is simple for them. Not only will you know where they are and where they're going, you'll also have a time stamp from each text (just in case things don't quite add up in the morning).

Sex + Texting = Sexting

Sexting or Sextexting is a terrifying concept for any parent, and it has become prevalent within the tween and teen community. What is it? Sexting involves inappropriate and/or nude photos of themselves or others that are texted to boyfriends, girlfriends, and even strangers.

Take this practice one step further to the brink of total devastation by adding the ability to forward and share these inappropriate photos with an unlimited number of people. Who knows where these pictures could end up? The possibilities are frighteningly endless.

"Not my child!" is what most parents think when the topic of sexting is brought up in conversation. The National Campaign (www.nationalcampaign. org), a national organization that seeks to improve the well-being of our youth, estimates that at a minimum, 33 percent of teenagers have admitted to sexting. The number increases to 36 percent when it comes to young adult women. Once more, the second someone hits the "send" button that includes a "hot pic" as an attachment to a text, serious consequences and severe harm are possible and can include the following consequences:

- Potential to be arrested and charged for child pornography
- Requirement to register as a sex offender
- Complete humiliation among peers
- Attractions of stalkers/sexual predators
- Hazing/name-calling
- Sexual harassment
- Possible expulsion from school

Teens have been arrested and booked for sexting, which in the eyes of the law can be considered a form of child pornography. Beyond that, sexting carries risk:

- Text Bullying — When compromising photos are texted to others in order to embarrass, make miserable, or destroy another person.

- Teen Pressure — Feeling pressured to text revealing photos—or else . . .

- Teen Suicide — Resulting from once private, inappropriate photos no longer remaining private, causing humiliation and devastation.

- Blackmail — Being made to do something against your will, because if you don't, that naked picture you texted of yourself will get sent to all your friends, teachers, coaches, parents, and anyone you would never want to see it.

● *Breakups/Relationship Problems* – Forced to stay in a relationship for fear of inappropriate photos surfacing somewhere unexpected.

Parents need to wise up and be aware that sexting is a reality, and it is becoming more common than we'd like to believe. We need to sit down with our kids to set texting boundaries and explain the potentially serious consequences to **ANY** kind of sex texting. Teens must understand how they could be exploiting themselves or others; that once a picture is attached to a text and sent, they can never get it back; and that the "love of their life" today could be the "jerk" of tomorrow who decides to forward the photo to all their friends. We must be sure they understand the serious consequences that arise from sexting and how it can negatively change their lives forever!

Sexting Prevention

One way you can prevent your teen from sexting is to request your service provider to restrict the sending of pictures with text messages. There shouldn't be a charge to eliminate this service, and your teens will still be able to text. They can still take photos, but they won't be able to share them with their friends.

There are software applications you can purchase to monitor all of your teen's cell phone activity including photos and text messages. Some of these software applications will also allow you to print a detailed report with all the information included. This is a feature that can come in handy if you ever need to submit information to authorities or school officials to prove a situation or report a problem.

When your child gets that most important of all gifts—his first cell phone—you'll need to have a sit-down to lay out clearly the dos and don'ts of proper phone use and set boundaries that work best for your family. Keeping the lines of communication open between you and your teen is one of the most important keys to parenting success and to your children making wise decisions.

Emergency 911

Texting can be a great benefit to your teens if they are ever caught in an emergency situation, because the technology associated with text messages offers additional benefits that regular phone calls can't:

- **Your teens might not be near a landline, but it's for sure that they're NEVER without their cell phones.**

- A text can be executed silently and instantly.
- Texts require less battery power than a wireless call. This may allow a text message to be sent when silence is critical to safety, even if the phone's battery is too low to make a call, landlines are down, or there's a low wireless signal.

Students with cell phones can quickly shoot off a text to multiple family members at the same time, instantly letting them all know where they are, that they're in trouble, and what their situation is. Having the ability to respond instantly to an urgent request for help is critical to both the sender and the receiver, no matter what the situation is, and text messaging provides the ability to do just that.

Last but not least, add ICE (In Case of Emergency) phone numbers, including area codes, into both your address book and your kids'. This way, you all have emergency contact numbers programmed into your cell phone. If you want to use text capabilities to report emergencies directly to the authorities, you'll need to contact your local police and fire departments to find out if they have added this technology to their service offerings.

Mass Memo

Most teens are loose lipped about pretty much everything these days. In the world of Generation

Text, all news travels fast, whether good or bad. Keeping a secret, and social events "by invitation only" are a thing of the past, creating even more texting pandemonium instantly. Any teen news can spread faster than a wildfire and to more people than you could ever imagine. Texting has unlimited boundaries, which can really be great, but it can also cause a lot of trouble.

Boost Your Text Reputation

If you already text, you may want to revamp and transform your text reputation! Don't send a text signed, "Love, Mom." Since your teens already have your name and number stored in their address books, they already know who's texting them. When they receive your text, the sender heading will automatically show **Mom Cell.** So, instead of using **Love, Mom,** consider adding **xoxo** or :).

Teens already find parents annoying in a zillion different ways, especially when their texts are too long and don't get to the point quickly. This is just another reason to work on your texting savvy, add in some text lingo, and send your note in an abbreviated format. Yes, it is possible to reduce your annoyance factor down to—maybe—just a million or so.

MYOB

Consider your teenagers' text messages to be personal and highly confidential, much like a diary or a personal journal. Not only would reading your kids' text messages be the quickest way to lose your teens' trust, but it's also a great way to misinterpret what your teens' texts are actually saying. When you're not the person sending or receiving the message, you have no idea under what context the message was sent. Was the message a joke? Was the message serious? Was it sent to be sarcastic? When you jump into the middle of a conversation without any background or knowledge about what was said, there's a good chance you'll misunderstand what you're reading.

As with anything else, always ask direct questions specific to a concern you may have rather than reading private information that could easily be misinterpreted and possibly shatter your teen's trust. If you're lacking complete trust in your teen, you can always limit or completely eliminate their text-messaging capability.

Text Addiction

Your teens probably have a difficult time being "in the moment" with the people they're with because of the constant text summons (that annoying ding or buzz that announces the arrival of another message).

It can be difficult to gain their full attention when their eyes are constantly looking down at their cell phones, reading their texts. It's pretty hard to be 100 percent present in face-to-face conversations, focus on homework, or hang out with the family when your teen's primary concern is his last forty-seven texts that he **MUST** respond to.

A parent's job is to set priorities and enforce texting boundaries, establishing how much texting is too much and making sure their teen's texting doesn't become an addiction—sending them straight into text therapy.

Take Charge, Create Text Balance

Many parents have a long list of pet peeves regarding texting, and they haven't been successful at enforcing "texts and balances" between just the right amount of texting and what is too much. Text messaging can be a great tool, but many teens have become addicted to their phones, the lifelines to their **ENTIRE** existence.

Parents should monitor texting activity, just like anything else that can be abused. In an effort to help out with this difficult task, wireless service providers offer parental control products, usually for a small monthly fee, that allow parents to set text boundaries. For example, AT&T Wireless can provide help with:

- Limiting the number of texts sent/received
- Restricting the times of day the cell phone can be used
- Controlling who can be texted (incoming or outgoing) by blocking and only allowing certain numbers
- Restricting pictures, video, or music
- Filtering content

Child Locator Services

The ability to locate a child is something most of us would pay any amount of money to have. Most wireless providers have options to add Child Locator Services as an additional benefit that you can add to your plan. One of the benefits of this service is that it allows you to locate a family member's cell phone via the Internet at any time, without a distance limitation (within the U. S.). Notifications of boundary breakdowns can also be set to notify you automatically by text message. Among other benefits, this service will allow you to pinpoint exactly where a cell phone is, as long as the cell phone is on.

GPS on your teens' cell phones can not only help keep track of your kids, but it can also save lives. It provides an incredible tracking device should your teenagers ever run away or be taken against their will.

Woo Hoo #3

Generation Text wants to be heard on their level, so go ahead and take the leap by entering into the digital era, and get on the same communication page as your kids. Learning how to text can open the lines of communication and help you keep up, stay informed, and be in the "know" of what's going on in your teen's everyday life.

Quick Text
Lingo Lesson

420 • Marijuana, Weed, or Pot

143 • I Love You

COT • Circle Of Trust

F2F • Face To Face

HOT PIC • Naked or
Compromising Photo

MINS • Minutes

P911 • Parent Alert

TBD • To Be Determined

TMI • Too Much Information

YW • You're Welcome

Hi TD&H! When R U getting off wk?
Lets meet 4 cocktails
@ 6:30-Sky High Lounge?

– Four –

From the Bedroom to the Boardroom

On a Friday afternoon, you know your man is in his weekly wrap-up meeting. You also know that this meeting won't stop him from returning your text—this is the one meeting he has each week when you know he can respond!

> Im out of here by 6. Sky High sounds perfect – just what I need! C U then! xoxo

Texts used in lieu of phone calls, e-mails, or love notes—which, **BTW**, are still really great—help you keep in touch. Texting allows you to reach out to someone at work, on the road, or at home to set or confirm plans, to let someone know you're thinking of them, or just to say hi.

Consider the text tool as another arrow in your bag of cupid tricks. For those who have just started dating, texts are totally private and can also be a seductive way to flirt. By being spontaneous and letting your favorite person know you're thinking of them, you will be sure to bring a smile to that special someone's face when they read your message. For those who are married, text messaging love notes and date night requests can add more va-va-voom into your va-va-marriage!

The Dating Game

Texting is a great tool to use prior to a first date. It can often help break the ice and point out basic likes and dislikes, which helps everyone feel more comfortable prior to that first, often awkward face-to-face encounter. Why not know ahead of time if your date is allergic to fish before making the reservation at that sushi restaurant you have been dying to try? Avoid swimming upstream and get to know the person you're about to take out. The more information you have about someone before you go out on that first date, the better off you'll be.

Guys and gals are very different—that whole Mars/Venus thing. Guys like being efficient, so they probably think it's great to ask a girl out for a first date by text. A gal, on the other hand, would prefer a more personal approach and could consider a "first date request by text" to be lazy and impersonal. She might even assume that a guy who texts a first-date request is the same kind of low-effort guy who won't open a car door or make a dinner reservation. Most gals prefer guys who are a little old-fashioned, polite, romantic, and personal—without being too wimpy. That being said: gals, go ahead and text your first-date request—guys don't mind being asked out by text message.

How about DNR LTR 2NITE?

Hot 4 U!

When it comes to your love life, now more than ever, text messages can really make relationship success much easier. Text messaging allows you to connect more often, when and wherever you want, even when you're really busy. It only takes a second to say: **"Missing U, 143, C U 4 Lunch,"** or maybe you're not sure what to say. You can say just a few words and if needed, read them nine times to be sure they're just right for the moment. :)

Texting allows guys and gals to be confident, funny, and romantic—traits that might be difficult to convey over a phone call or in person when a relationship is just beginning. Take the shy guy who might not be comfortable saying, "Wow, you are an incredible kisser. I had an amazing time last night!" Or, for the guy who's trying to send a negative message, "I'm moving to Alaska tomorrow, so I won't be able to go to that party w/u." Texting can make this type of exchange easier, even if the message isn't exactly what you were hoping to hear.

Texting definitely has advantages when dating. You can connect with that special someone more often and strengthen bonds by touching base several times a day. But wait—before you send that cute little text to a potential love interest, think about how to send just the RIGHT message.

Flirty Texting

Sending a text message to someone you're interested in and have had prior communications with is a great way to get to know someone better. It's amazing how much communication opens up when text messages are being exchanged back and forth with someone of interest. Throw in a text lingo from the lingo dictionary here and there as a way to flirt. It's fun when you have the upper hand on the meaning of your abbreviation, and they have no idea what you're saying, but they really want to know! Your flirting by text can be just the thing to move your relationship in the right direction.

> R U free 2nite?

However, texting an out-of-the-blue message like this one could be a pickup line that brings you success, or it could be a total text bomb. It all depends on your recipient and how well you know each other.

Booty Beware

Use your best judgment when you get a "Booty Beware" text at midnight, wanting to know what you're up to and if you're available. And believe me, even though it's a text, you can still hear that slurrrr through a text or be quick to identify when a comment makes absolutely no sense whatsoever. If it doesn't make any sense, trust your instincts and consider letting it go unanswered. If you're lucky, your sender will forget he sent it to you in the first place. If you are up for a booty text from your "friend with benefits," then you know what to do!

Lightning Fast

When exchanging texts, it's not the best time to play hard to get! Whenever possible, don't delay your text response to that special someone, because the great majority of texters return messages as fast as lightning strikes. A delay when replying to a text, unlike an e-mail or a call, could be perceived as if you don't care or even as a complete blow off.

If you need to wait to text back because you're on the phone, in a meeting, or just can't be interrupted for whatever reason, be sure to explain the reason for your delay, so your SO (significant other) knows it wasn't intentional and you're still "feeling the love."

In a nutshell, if you're going to commit to text messaging, you need to try to respond quickly whenever possible. You'll need to be on top of your text game. By paying attention to your texts and practicing, you'll move up to the premier team of texperts in no time!

Texting in a Man's World

Text messaging doesn't discriminate between the sexes; everyone loves the technology because it's quick, easy, and private. Guys are the masters of solving problems and getting what they want. They also have their own reasons that explain why texting is a great and handy tool they like to keep in their communication shed. Let's be honest, can a guy ever have too many tools?

Here Are a Few Reasons Why Guys Text:

- Perhaps he's interested and thinks texting will work to his best interests.

- Maybe he doesn't have the courage to call.

- He doesn't have time to call, but can easily send a quick text.

- Minimal words and time are needed, since texting is short and to the point.

- Asking personal questions is easier, as is providing personal 411.

- He could be texting seven girls at once. :(

- Guys may believe that texting, as opposed to a phone call, keeps them connected but at a safe distance.

- They can say "I'm sorry" or "I was wrong" more easily.

When your man texts you, consider it a compliment that he cares enough to send a personal note. One thing's for sure—there's nothing sexier than a guy who can communicate, especially if he can send a quick text right in the middle of his busy day to say, thnking of u, cnt wait til 2nite!

Just remember to keep the mundane or silly texts away from your guy—and don't get too offended if he doesn't respond to all of your messages. Sometimes their big fingers don't hit the right keys, and sometimes they get tired of texting and would rather talk. You'll even find that when your guy is tied up at work, multitasking (working and texting) isn't his strongest suit, so he doesn't respond. Either way, as long as you both understand the way each of you prefers to communicate, you'll be halfway to relationship success.

Only in a Woman's World

Women are the masters of multitasking and the experts at multilevel ball juggling; they meet stress with sass and hardship with humor, no matter what time of day or night. Texting offers gals one more way to connect with friends and family, while still managing the seventeen other tasks at hand that still need to be completed.

- ☺ **Gals don't like to say no. Texting helps us manage our over-committed commitments.**

- ☺ **Gals are overachievers; texting helps us achieve more.**

- ☺ **Gals like to be in the loop with everyone, all of the time. Texting helps us do that!**

- ☺ **Gals love to gab and socialize. Texting is another tool that helps us do more of that.**

- ☺ **Multitasking is one of our claims to fame as gals. Texting affords us a simple way to communicate with seven people at the same time.**

One of a gal's most coveted assets is her girlfriends. Day or night, no matter what time it is, women's friendships are a priority, and texting helps them stay close. **"I've just been fired—can we meet??"** or **"OMG, your X just walked in to Apple,"** or **"I'm @ the mall—get down here!"** Friends are always there for each other, no matter what!

Remember to RBTL (Read Between the Lines)

Pay attention and listen up! You need to **RBTL** when you're using text messages to communicate, so pay special attention to the TM red flags that could mean "he's just not that into you:"

- When the text you just got doesn't quite add up or make perfect sense.
- When he only texts and never wants to talk face to face.
- When your entire relationship exists over text messaging.
- When texts always come late at night and hardly ever during the day.
- When responses to your texts are always delayed.

Take notice and consider these red flags when they occur; take the hint and don't waste your time waiting for the next text delay or no response text.

> Sorry, I was in meetings all day.
> I've been busy drinking beer with my buds.
> Didn't get your txt til I got home.
> Maybe we can hook up @ midnight.

If this sounds familiar, and happens all the time, it's probably time to kick 'im to the curb and look for someone who will appreciate you for who you really are.

In the Vault

Unlike e-mails, text messages are totally private and don't leave a trace for anyone else to read (unless it's intentional, or a cell phone breach occurs). Text messages can assist you in making that love connection you've always dreamed of, leaving you flying high on the jet stream of life.

Woo Hoo #4

Texting can be a great way to augment your relationship with that special someone or help keep the flame burning. You can also go for the flirty-dirty text message if you think your lovah will appreciate it—just understand the risks involved.

Quick Text Lingo Lesson

BCNU · Be Seeing You

H&K · Hugs & Kisses

ITS A D8! · It's A Date

JMBA · Jamba Juice

MUSM · Miss You So Much

RBTL · Read Between The Lines

SWAK · Sealed With A Kiss

TD&H · Tall, Dark, & Handsome

TLC · Tender Loving Care

WYWH · Wish You Were Here

XOXO · Hugs & Kisses

**411 Info: Neiman Marcus,
1618 Main Street, Dallas, TX
214-741-6911. Click below for map.**

- Five -

Just Text It: Go Green, Text Smart

You're at a neighborhood café for a quick afternoon Mocha Frappuccino, extra whip, when a good-looking man taps you on the shoulder and asks you for directions to Neiman Marcus. Because of your text savvy, you dial 411, ask for Neiman's, and then select the option that allows the listed information to be texted to your phone. Within twenty seconds, you have all of the pertinent information on your cell phone, including a map with the driving directions and the phone number in case he gets lost. The guy's pretty impressed and asks you to forward it to his phone. Now he has your number and you have his. Fabulous!

Information at Your Fingertips

For life's little reminders, there's an unlimited number of ways you can use texting that work hand-in-hand with thousands of different Web services, offering you many ways to get the most of the information you need, texted straight to your cell phone, and saving a few trees all at the same time. Here are a few more things you can do to get the most from your TMing ability:

- **Check Flight Status**
- **Track a Package**
- **Get Headline News, Local News, and Weather**
- **Receive Financial and Stock Information**
- **Track a Person, En Route and Traveling**
- **Shop and Play Texting Games**
- **Get Driving Directions or Have an Address Texted to You**
- **Access 411 Information**
- **Text Bill O'Reilly on** *The O'Reilly Factor*
- **Make Reservations**
- **Get Boarding Passes**
- **Enhance Your Marketing Efforts**
- **Send Emergency Requests or Ask for Help**
- **Report Police Tips/Crimes**

- Expand Your Networking
- Generate Medication Reminders

Frequent Flyers

Text messages work perfectly for the frequent traveler. Travelers can eliminate the need to carry paper confirmations because all pertinent reservation information can be texted right to their cell phones.

- **Major airlines may offer to text message flight status notifications.**
- **Major hotels may offer to text message confirmation information.**
- **Online mapping services may offer to text directions.**
- **411 information**

Help save the earth and "Go Green" by taking an extra second to request whatever information you're needing be texted to your phone. You'll eliminate the need to print your reservations, confirmations, and maps. You'll not only become more efficient, you'll also be doing a small part to help improve our environment.

Text Your Vote

How many times have you called to vote for your favorite TV singer or dancer, only to get that annoying busy signal? Text your votes instead and never get a busy signal. (This benefit depends on your cell phone provider, which will be mentioned on each TV program that offers the "vote by text" option.)

- **Dial the number you are asked to text your vote to.**
- **Press "send" as if you're going to call, let the call connect, and then hang up.**
- **Now, select the number you just called and select "text/SMS."**
- **Then hit "send" again.**

That's all there is to it! You've just voted, without the risk of receiving a busy signal and without having to hit the redial button over and over. Because of the technology text messaging uses, your message won't get rejected because the landlines are busy. Instead, your message waits in line to be delivered, leaving you free to move on to your next big adventure.

Party Animals

Vote4me! According to The Neilson Company, during the Obama campaign, 2.9 million Americans received the text message announcing Joe Biden as Obama's running mate for the 2008 election. At that time, this single message was said to be the largest announcement sent by text message in the U. S.

Texting is changing politics dramatically and globally, and it's transforming the way we receive political information. TMing provides the ability to put compelling and inspirational messages right into the palm of another's hand. Maybe you've already seen text politics in action?

PWR 2 the PPL

Here Are Just a Few of the Most Common Uses:

- Fund Raising Messages
- Campaign Promises and Announcements
- Organizing of Events, Including Protests
- Reminders to Vote
- Election Results
- Disagreements and Political Bashing
- Support Cause Messages
- Requests to Post Your Opinion

Text messaging, Twitter, and Facebook are all powerful, influential, and cost-effective forces used around the world. They can have a huge impact on a country's politics, influencing voter turnout and affecting election results. Can you just feel the power?

Crime Watch

Police departments across the country are trying to keep pace with communication technology to maximize the ways people can provide potentially valuable information about a crime they've witnessed.

As reported by MSNBC, special software has been designed specifically for police department use for exactly this reason. It protects the tipster's identity by sending the original message through a third-party service that scrambles the phone number and then sends the message directly to the police department. Police can text back the tipster if they choose, as the original message is given a random code that connects to the sender's cell phone. Contact your local law enforcement agency or go online to determine if this service has been made available in your area. Chances are, if the service isn't currently available, it soon will be.

A growing number of cities around the United States are using text messaging to report emergencies and anonymous crime tips. Many are in the process of expanding the options of how you can report crimes, hoping to increase the number of anonymous phone tips by accepting text messages from the ever-expanding generation of texters.

Event Texting

During the Super Bowl, according to the *New York Times*, the NFL implemented a text message hotline for fans to alert security anonymously about unruly behavior. The goal was to enable fans to report drunken neighbors without confrontation, and it worked!

Tween and Teen concert promoters are jumping on the text band wagon too. They set up receiving stations for parents to retrieve their kids at the end of a concert when signaled by a text message.

Woo Hoo #5

There are a ZILLON ways to interact by text—and now that you know just a few of them, you'll start noticing the text option everywhere you look. Think of using text notifications as a substitute for that dream personal assistant you've always wanted but haven't been able to afford or gotten around to hiring yet. True, technology can't replace a person, but it's a great resource you can easily use to your advantage and still have some extra cash left over to go shopping!

Quick Text Lingo Lesson

411 • Information

747 • Let's Fly

BIZ • Business

CONF # • Confirmation Number

CRLTN • The Ritz Carlton

G5 • Gulfstream V

JTLS • Jetless

NP • No Problem

RESO • Reservation RTZ

V-AIR • Virgin Airlines

SWA • Southwest Airlines

– Reminder –
Chapter 3 Test Tomorrow

– Six –

Text Benefits for Anyone and Everyone

Thankfully your eighth-grade daughter just got this message on Sunday morning from Mr. Melby, her chemistry teacher. Mr. Melby's in touch with his students and likes to send group text reminders for upcoming tests or due projects. Since Mr. Melby started his little reminders, he's seen his students' grades rise and his reputation and respect soar.

Bringing technology into the classroom to get kids more interested and enthusiastic can excite them about learning. For example, text messaging can help teachers keep students informed by providing homework assignments to their students who miss class. Some teachers even text students over the weekend to remind them about a paper that's due or a test that's occurring on Monday. While most teachers prefer to have the cell phones turned off during class time, high school teachers and college professors have found texting to come in handy for reminders outside of the class.

Wired in the Classroom?

Yep, it's happening. Today, kids relate and use cell phones in a different way than their teachers and parents do. Yes, they text message their friends enough to drive everyone around them crazy, but their reliance on their cell phones within the classroom goes far beyond social networking. Students also use their cell phones to:

- Snap photos of the homework board instead of writing it down.
- Go online to check their current grades.
- Store friends' and families' information in their digital address book.
- Check the time instead of wearing a watch.
- Fill their calendars with quiz and test dates, school events, and after-school activities, such as team practices and club meetings.

Lecture Texting?

According to the British Council, "It's not just students who are texting, professors have also realized the benefits." For excample, the staff at Wolverhampton University is now sending students revision tips, timetables, appointment times, and coursework feedback using text messaging. Some of the reasons colleges are finding value in text messaging include:

- Accessibility ~ Students can be easily reached by the staff, and vice versa.
- Private ~ Texts are private and do not violate privacy laws.
- Control ~ Provides professors unprecedented control over how they communicate with their students.
- Encourages Participation ~ Students participate more when they're not put on the spot in front of their peers.

Lecture Software Encourages Learning

Some professors are using new software to encourage their students to participate more during lectures. Dr. Kevin Linch, a professor at the University of Leeds, uses such software. This is how it works. Students can text questions to a predesignated number, which then gets projected onto a giant screen for everyone in the class to see. The students then have the option to text in their opinions or answers, or the professor can simply answer the question verbally. Dr. Linch commented that "texting answers have two benefits—multiple results come up live on the screen, and texting encourages responses from some students who might not want to put their hands up."

Teachers who use popular technology like texting can encourage learning and increase participation from their students. Most students of all ages are connected. Continuing their connection inside the classrooms and lecture halls can be very beneficial to teachers, staff, and students.

Get in the Loop

Some schools are using texting to keep parents in the loop on their children's progress—praising positive attitudes or addressing the need to work out behavioral problems. Parents are welcoming the additional communication because they can be told what's going on, wherever they are, at the very moment a particular problem is occurring.

If a parent needs to come to school immediately, a text message is the fastest way to deliver a message:

Brandon needs to be picked up ASAP.
He has a fever of 101°.

To Lingo or Not to Lingo

Most educators use lingo in moderation, but they have the common goal of communicating with students in the method they prefer. Because teachers set the standards of behavior, maturity, and commitment for their students, sticking with the basics may be the best idea. Every generation creates its own language, attempting to keep the adult world clueless, and teachers are savvy enough to want to be part of the texting phenomenon. Sending or receiving a text with grammar and spelling completely intact still says a teacher is tuned in.

Student MIA

Administrators and teachers are utilizing text messaging to contact parents when students don't show up for school (MIA = Missing in Action). Texting is usually preferred over a phone call because of the cost and time savings it offers schools and because of its effectiveness. Rather than leaving a voice mail on a home phone, which a parent might not get until much later in the day, texting provides instant and direct communication. Many families are even canceling home phone services, so texting to cell phones becomes even more important and necessary.

Jennifer is not in class today. Is she sick?
If so, remember to call into the
office to excuse her.

Campus Alerts

Campus emergencies and mass notifications that need to be delivered immediately can be sent using text messaging, which delivers urgent and important information directly to a parent's cell phone. Messages can almost be guaranteed a quick delivery, without encountering busy or dropped signals. Think about catastrophic situations like a campus fire, a shooting, an intruder, a flood, or an earthquake, when landlines aren't dependable. Text message to the rescue!

> 911- 911- 911 – Campus on Lockdown.
> Do NOT Come to School. Update to Follow.

Coach's Corner

Coaches love to use texting to change or cancel practices and to send important information to parents and team members, because it's quick and doesn't waste time. If you're a coach, texting is a great tool to notify players and parents of a change of field or that the start time for Saturday's game has changed. Texting is also a great way to communicate important information to parents, helping to avoid miscommunications from players or important information from getting lost in translation. TMing eliminates the need to make separate, individual calls, which is time-consuming and inefficient when compared to a multirecipient text. If practice is cancelled and the notification is texted, changing plans is so much easier for everyone involved.

> Practice Time Changed to 5:00 on Field 8

Privacy Mode

Unlist your phone number! Sometimes privacy, especially for underage tweens and teens, is a concern for all parents. Some don't want teachers or coaches sending individual texts to their child's cell phone, and some teachers don't want thirty fifteen-year-old kids texting replies to a simple text notification.

Direct cell-to-cell text communication can be avoided by posting comments and announcements to a publicly accessible Web site like Facebook, MySpace, or Twitter, which allows students to "opt in," enabling the delivery of text messages instantly to their cell phones.

Another alternative for mass text communication that provides privacy to the sender is submitting text messages to a centralized mailbox residing on the Internet, which then routes your message to a distribution list. Teachers', coaches', students', or players' cell phone numbers on the distribution list are invisible to all parties.

These distribution options provide privacy to teachers, coaches, students, parents, and players and can also provide an audit trail of all text communication that's been sent and received. In general, there is full accountability for everyone involved.

Technology is driving change globally, at home, at work, and at school. Students will not only think they have "one hip teach," they'll also be more apt to engage and learn what's being taught to them.

Aging Up but Not Out

If you're a grandparent, you may be wondering what happened to those darling little cherubs who used to love to stay the night and hear you read out loud to them as they helped you turn the pages. In fact, they used to love doing just about anything with you, as long as you were together. When you called them, they would talk, giggle, and laugh with you and tell you about their day.

Where did those kids go? If your grandkids have morphed into teenagers, you're probably finding that your phone calls don't get returned and your e-mails seem to disappear into outer space.

If you're wondering why you never get responses from all those e-mails you send or calls you make, **please pay close attention and take serious note:** for the most part, teens don't e-mail like their parents and grandparents do. If your calls are going unanswered, it's because most kids don't let their cell phone ring (it's on vibrate), and they don't check voice mail for messages. The only way you're going to get their attention is to text them.

When you make the decision to not let new technologies hinder you or keep you out of the communication loop, go ahead and send your first text to your favorite teen. (Refer to the Quick Start Guide in the beginning of this book for simple instructions on how to send a text.) **This comes with a warning:** please be prepared to get a very quick response that says, "**Who's this?**" Just understand that there's a good chance they don't have your cell phone number in their address book, so they won't know who's texting them. Just laugh to yourself, and be happy that your text generated an immediate response.

Texting is a wonderful way for grandparents to stay connected to Generation Text; it allows you to easily communicate across generations. It's certainly not the same as hearing their voices on the phone, but something is better than nothing, and with texting, you'll be able to reach out more often than you would with phone calls. You can say a lot in 160 characters, and you can do it as often as you'd like:

- **How did your chem test go? I'm sure you did great!**

- **Did u win your game today? What was the final score?**

- How are you feeling? Your mom said u were sick...I hope u feel better!
- When can we get together and see a movie? My treat!
- When are finals over? We need to celebrate!
- Just thinking of you...xoxo

Since texting knows no boundaries, you can stay in touch around the world, no matter where your children, grandchildren, and friends are living. What could be better than that? Yes, we're all aging up, but that doesn't mean we have to age out of the day-to-day activities of our family and friends.

Deaf and Speech Impaired

> Dad, Can I go 2 Danielles aftr schl
> 2 stdy & hv dnr?

Paige just finished school, and she wants to go to Danielle's house to study for an exam and eat dinner with Danielle's family. Paige's dad is deaf, so she utilizes text messaging for all of her communication with her father when she is away from home.

- For the hard of hearing and speech impaired, where traditional communication can often be difficult, text messaging is helping to remove linguistic barriers and is providing an additional sense of independence and freedom within this community. In the U. K., 98 percent of hard of hearing people use text messaging daily as one of their primary modes of communication (Birmingham Institute of the Deaf).

⊘ More than thirty million American citizens have speech or hearing impairments (National Institution on Deafness and Other Communication Disorders–NIDCD).

⊘ According to the World Health Organization (WHO), there are 278 million people who have moderate to profound hearing loss in both ears.

⊘ The ease, speed, and portability that text messaging offers are key benefits to the hard of hearing.

⊘ Texting liberates the deaf community from being tied to a TTY (a specialized text telephone).

Plans for the Hard of Hearing

For the hearing or voice impaired, some cell phone providers offer plans that don't charge for the basic voice plan and offer unlimited text and data capabilities. This is usually cost effective because you don't pay for all of the charges associated with voice services.

No-Cost Texting

The deaf and hard of hearing can use a no-cost text messaging service at www.text4deaf.com. This service makes it simple to send and receive up to ten free text messages a month from your computer. Text4deaf also offers a premium service with unlimited texting, among other things, for a nominal monthly fee.

Video Calls

Add a video call to your communication arsenal. The combination of video calls, IM, and texting enable families and friends to keep in touch more often.

Real Time Text (RTT)

To take text technology one giant step further, the use of Real Time Text (RTT) is even more dynamic and interactive than text messaging. RTT can still be done on cell phones, but two conversation boxes are displayed on the screen, one for each user. One person types a message, while at the same time the other person sees the words forming, letter by letter, in one of the text boxes. In the other box, you can write responses, ask questions, and even interrupt—all at the same time. Real Time Text makes it possible for the hearing or speech impaired to participate in natural, direct conversations that are a lot like verbal conversations.

Text messaging and Real Time Text are incredible tools for the hearing and speech impaired, helping to provide comfort and security in the event of an emergency, as well as assistance with day-to-day communications. Texting also helps promote independence and breaks down some of the communication barriers between the hearing and the hard of hearing.

Emergencies and Urgencies

Having the ability to text 911 is a tremendous emergency service that's offered by more and more American cities in an attempt to broaden services to the speech and hard of hearing. Texting provides the deaf a potentially lifesaving tool to report crimes and request police, fire, or medical assistance by utilizing text messaging as an alternative to the telephone.

Many major cities around the world have implemented internal processes that allow emergencies and crimes to be reported by text message. Without an easy way of communicating with emergency services, the hard of hearing have limited options. In the past, they had to send text messages to friends or family and hope someone called for help on their behalf.

The use of text technology makes this situation much easier as well as simplifying many other types of interactions that many of us take for granted.

Woo Hoo #6

The use of text messaging by schools, educators, and coaches is becoming more and more popular as text messaging becomes more prevalent within everyone's day-to-day mode of operation. As for the speech impaired and hard of hearing, they've been texting as a way to communicate for over a decade. As more and more of the general population utilize text messaging, more services are becoming available for everyone's benefit.

Quick Text Lingo Lesson

2NITE · Tonight

B/C · Because

BYAM · Between You And Me

CHG · Change

CME · Crack Me Up

ERLY · Early

IMO · In My Opinion

L8R · Later

MYHOTY · My Hat's Off To You

PEEPS · People

P/U · Pick Up

Hi Jesse, can u qukly txt me David's cell phone #? I've been waiting @ the restaurant 4 30 mins now...thx

– Seven –

Corporate Texting– From Main Street to Wall Street

Your salesperson is sitting at a restaurant, waiting for her client to show. She's quickly and quietly able to send a text to her personal assistant without bothering her nearby diners. Jesse quickly replies with the client's cell phone number. Your salesperson sends a text to her client, confirming she has the right time and location for their meeting. Due to the nature of texting, your employee has been able to rapidly solve a typical business problem, while at the same time staying within restaurant phone-etiquette boundaries.

Whether your company is on Main Street or Wall Street, businesses are thinking outside the box by taking advantage of what text communication offers. Text messaging is a significantly less expensive way to reach customers, as opposed to other media options, such as TV, newspapers, radio, and direct mail.

Have you received a TM from a financial institution, retailer, salon, utility company, healthcare provider, or campaign headquarters? Just look at how texting changed our forty-fourth president's campaign! In 2009, texting played a significant part in the election process. President Obama was the first American President to have highlights of his inaugural speech texted live to anyone who registered under www.America.gov. That's one hip prez!

Trending Statistics

According to www.Textually.org, some of the key trends that point to why companies are investing in text innovation to attract new customers include:

- **Text messaging is growing at a triple-digit pace.**

- **Four million landlines are shut off every year, and pay phones are quickly disappearing.**

- **There are over 1.5 billion Internet users worldwide.**

- There are over 4 billion cell phone users worldwide.
- Americans send more text messages than they make phone calls.

> Congratulations for being our 1,000,001 customer! You've just won $500 towards the purchase of anything in our store!

Take a look at a few ways text messaging is being used by both consumers and companies:

- Information alerts for news, weather, stocks, and finance
- Notification of new services and products
- Sales and marketing
- Appointment reminders
- Sale notifications
- Customer service/power outage notifications
- Notification of product recall
- Location-based notifications to customers in a particular area
- Medication reminders
- Test result notifications
- Banking information
- Payment processing
- Sports team updates
- Election information

Business Benefits

> Your appointment has been cancelled . . .
> the doctor has had an emergency and won't
> be in the office today.

Having the ability to embrace new technology is one of the keys to a company's success. Mobile marketing unlocks your business's capabilities to communicate globally, providing the ability to increase communication—a top priority for virtually all companies.

Here are some of the significant advantages text technology offers when compared to traditional marketing and advertising:

1. *Cost* ⟿ Text messaging is far less expensive than most other mediums.

2. *Personal* ⟿ When text messages are sent and received, they feel unique and personal, as if the message was delivered to you, and only you—like a private note, for your eyes only.

3. *Immediate & Targeted* ⟿ TMing is more immediate than e-mail and more targeted than television, newspapers, or radio. It allows for companies to reach millions of customers quickly, efficiently, and effectively.

4.   **Trust** ~ Test messages are a trusted method of communication, as most TMs come from people or companies you know.

5.   **Generation Text** ~ There's no better way to reach this group of consumers, since texting is the preferred mode of communication for Gen Text, one of the most coveted demographic groups targeted by marketing and advertising campaigns.

6.   **Spam Reduction** ~ Spammers target e-mail addresses at a rate of more than 110 billion e-mails a day. On the other hand, wireless service providers heavily protect and monitor their networks from spam invasion and shut down spammers as soon as they're first detected.

7.   **Location Specific** ~ Are we ever (willingly) without our cell phones? They are with us almost all the time. Using GPS technology, advertising, or information text messages can be targeted to customers who are in close proximity to your establishment.

Text messaging is a fast, inexpensive, and effective form of mass communication that can be target marketed to your customers quickly and inexpensively. Be sure to eliminate text lingo when contacting customers, and make the message short, sweet, and to the point:

> Visit your local Apple Store this weekend
> and receive a free iTunes Download

Customer Contact

Text-messaging communication is multifaceted and can provide significant benefits to customers and to your bottom line. For example, TMs can be used to remind customers of upcoming appointments; it's a great way to announce a new product or a product enhancement or to send important notifications. Can you spell SALE?

> Summer blowout @ Target!
> 50% off all patio furniture and BBQs

By building simple integration to a Short Message Service (text) gateway, a CRM (Client Relationship Management) system can easily and automatically be scheduled to send a text message as a reminder of an upcoming appointment or a mass notification to a large group of customers all at once. If you're considering adding text messaging as part of your appointment reminder system, you should use software that integrates your reminders into your company's internal software application system.

Also be sure to have your customers "opt in" to any TMing programs you initiate, because your clients' cell phone providers may charge individually for each text they receive. You should also examine privacy policies (such as HIPPA) that might be applicable to your industry when electing to implement new communication alternatives.

Don't overdo your text message communications to your customers. Keep your messages important to the recipient, and don't send too many messages too often. You don't want to be considered a spammer, only to have all of your texts deleted before they are read, or worse . . . have your customers ask to be removed from your text delivery service.

Corporate Text Plans

When texting on company time, you'll need to get permission from your employer or verify that your company has added a text program to the corporate plan before you start sending and receiving text messages on any company cell phone.

Pay close attention to whom and how often you're texting, especially when you're texting on company cell phones, during company time. Because policies and procedures can range drastically from company to company, check with your HR department to find out your firm's stance on text messaging.

Texting Associates and Coworkers

For shoptalk, texting can help coworkers keep in touch whenever they're out of the office or not close by. Just keep all communication above board and stick to the task at hand—building your business and turning a profit. Be aware that your harmless flirtations with that hipster from accounting could heat up to a full boil as a result of your flirty texts. Let your inner-you integrity be your guide to appropriateness. Consider sending your personal messages from your personal cell phone.

> Heads Up - Boss is on way down and he's pissed, have u finished the 9th draft?

In Meetings

Appropriate professional etiquette dictates that you shouldn't sit in a meeting with your cell phone on the conference table or in view of your colleagues, coworkers, and clients. Everyone will be able to see the light blink or hear the vibration from every e-mail, call, or text that comes in. Just in case you possess the text-under-the-table talent, and you think you'll get away with texting during meetings, think again—everyone knows what you're really doing under there. The no eye contact and flying thumbs are a dead giveaway!

Text messaging can be a handy tool while you're on a conference call. If you want your coworker who's giving the presentation to mention something important he forgot, but you don't want the other attendees to know, send him a quick text:

Don't 4get to go over pricing and rev share.

Text messaging can be a great tool to use in order to connect with coworkers. Just follow your company's guidelines and texting etiquette, and you'll be on your way to becoming a professional texter!

Text Lingo @ Work?

To lingo or not to lingo will depend on what type of business you're in and your audience. Use your best judgment, especially when communicating with your customers. Many companies have implemented a strict "no text lingo" policy when connecting with customers and for client notifications and advertisements. You know your customers better than anyone, so use your intuition to do what's best for them.

Text Documentation

If you are using text message technology on your PC or Mac at the office, conversation logs of text messages could be routing through your corporate servers, which would place an imprint of your message within your company's network.

If you're texting on your cell phone, whether company owned or personally owned, most likely your conversations won't be logged. But on most cell phone statements, providers will detail each text message sent and received by date, time, and cell phone number. Check with your IT professional to get the DL (down-low) as to how texting works within your systems. Again, use only your personal cell for personal texting.

Woo Hoo #7

Companies and organizations can benefit from corporate text messaging programs. Plans are typically cost effective, quick, and easy to execute. Most customers like text messages because they are fast, private, and less intrusive than phone calls.

Consider investing in text technology as just another option to market/advertise your products or services, send reminders and notifications, or communicate financial information. Contact your IT professionals for recommendations that are specific to your company or organization's individual needs.

Quick Text Lingo Lesson

DD • Due Diligence

DQYDJ • Don't Quit Your Day Job

EMBM • Early Morning
Business Meeting

GMTA • Great Minds Think Alike

HIOOC • Help, I'm Out of Coffee

KISS • Keep It Simple Stupid

MSFT • Microsoft

NWR • Not Work Related

OTP • On The Phone

P&C • Personal & Confidential

PIA • Pain In The Ass

STD • Seal The Deal

SWAG • Scientific Wild Ass Guess

Wish u were here...
its not the same w/o u!

– Eight –

Texting Tips and Tricks for the Text Savvy

You were supposed to go to New York City with Wendy, your BFF (best friend forever). At the last minute, you had an emergency and couldn't go. You're sitting in the hospital with your dad, hoping for the best, when this text message arrives. You smile and send back the following message: **I wish I was there 2, I'd have much more fun w/u– that's 4 sure. B sure 2 Have a Cosmo 4 me! :).** Wendy texts back, **U got it, we will b back in NYC 2gether next year! :)**

Text messaging might already be a part of your life, but maybe you're not sure how to text a picture or a short video. Maybe you're not sure what text plans are available, or you're not positive how text messaging works when you travel abroad. This chapter provides you with some helpful hints, tricks, and texting tips.

Picture Perfect, Video, and Voice!

A picture is worth a thousand words, and since you're now a texpert, let's elevate you to the next level and give picture and video messaging a try.

Multimedia messaging works when you want to text a picture, video, voice, or audio, along with your words. It's easy. Just snap a photo or shoot a short video using your phone's embedded camera and send it to any compatible phone by addressing it to a wireless ten-digit number or e-mail address.

Picture, video, and voice messaging allows you to snap a picture and send it along with a quick text message to anyone with a cell phone, just like that! Share your favorite moments with family and friends wherever you are with video messaging. It takes only between 25–30 seconds to send, though it can take a few minutes to receive.

Here Are Some Important Tips from AT&T:

- Picture and video messages are priced differently from text messages.

- You'll need to add picture and video messaging onto your plan.

- Send short video clips and lower resolution pictures—try to stay under 600 KB.

- You'll usually be charged for each picture or video you send and receive, unless you've purchased a data package that includes this feature.

- If you purchase a picture/video messaging plan, you're allotted a number of messages that are priced lower than the per-message charge.

- When a single picture/video message is sent to multiple recipients, the sender is charged for each recipient, and each recipient is also charged.

- There's usually no additional charge for sending a picture/video message while roaming within the United States.

- When sending or receiving international picture and video messages, check with your provider regarding how these are priced and how it works.

Verizon Offers These Simple Steps to Text a Picture or Video Message:

1. Select the person you want to send a photo or image to, and instead of hitting the "send" button to make the voice call, select the MMS option to their ten-digit cell phone number.

2. The number will appear. Then hit your menu button again and select "attach picture, video, or audio."

3. Select where you want the item you are attaching to come from, usually device memory.

4. Select which photo, video, or audio you want to send.

5. Hit the send button.

> Chck this out! Found on sale!
> Do u want a pair?

Group Texting

As long as your message is short and sweet, sending a text to a group of recipients can be handy when appropriate. Here's how to do it:

1. Select the first recipient of your text.

2. Select your "menu" option, then select "add recipient" or "add address."

3. Your address book will come up, and you can add the cell phone number of the person you want to include in your group text.

4. Hit send.

Sending group text messages isn't normally the best idea, unless you have a special occasion to announce to a big group:

> Happy New Year!
>
> Just Married – Pat & Patrick – August 27th!
>
> It's a Healthy Baby Boy! 7 Lbs, 9 Inches (Picture Attached)

One of the benefits of a group text is that recipients can't tell who else got the same message, so they feel as if your message is personal. Just be sure to consider to whom you're sending your message, so you can stay cool and keep it fresh.

Forward . . . Text

With text messaging, you can easily forward a message to someone else within seconds. Here's how to do it:

1. Select the message you want to send.
2. Select your "menu" option, then select "forward" or "forward as."
3. Your address book will come up, and you can simply add the cell phone number you want to forward the text to.
4. Hit "send."

It's important to know that the message you forward comes up exactly as it was sent to you. If you need to make any changes or additions, you can do so right before you hit send. You'll just have to identify where the original message ends, and where your part of the message begins. Try adding several spaces, or just return down to the next line for easy reading.

No Read Receipt

Unlike e-mail, text messages don't have the option to add a read receipt. You can get delivery reports on most cell phones, but they will only confirm the message was delivered to the recipient's phone, not that the message was ever read. A quick response is the best read receipt for a text message. Normally, if you don't get a quick response from your text, either your textee accidently skipped over your message or they haven't had a chance to read it yet.

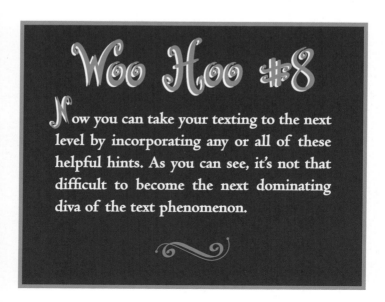

Woo Hoo #8

Now you can take your texting to the next level by incorporating any or all of these helpful hints. As you can see, it's not that difficult to become the next dominating diva of the text phenomenon.

Quick Text Lingo Lesson

A2D • Agree To Disagree

AFAIC • As Far As I'm Concerned

BAU • Business As Usual

CX • Cancelled

DBL8 • Don't Be Late

FWD • Forward

GGL • Google

L/M • Left Message

TWD • Texting While Driving

TWTR • Twitter

WTG • Way To Go

-Nine-

Text Your Tweets to Stardom

Twitter will ask you only one question: "What are you doing?" As long as your answer is 140 characters or less, it's possible for thousands, possibly millions of people (if you're a Twitter Rock Star) to see your answer the moment you hit "enter." Stardom, here you come!

Twitter is a trendy communication tool that's part micro-blog, part social network, part instant message, and part text message that allows you to share information by sending quick updates in 140-character snippets or less. When using Twitter, you'll be sending short updates ("tweets") to your community and passing on what you learn from others to friends, family, colleagues, and anyone else who chooses to follow you on Twitter. When you send tweets, they'll be posted directly to your Twitter site, which is then automatically sent out to all of your followers. Your tweets are searchable and visible on Twitter Search. Twitter is social messaging meets Facebook, at its best.

Twitter Benefits

One of the benefits of Twitter is that you don't have to keep up with another Web site if you don't want to. Just set up your account so all your incoming updates go to your e-mail address or your cell phone as a text. This way you can read them when they come in and respond when you want.

Tweet What?

You can have fun sending updates that include attachments of your favorite articles, songs, and important news that you'd like to spread, or you can just tweet something fun that tells your followers "what you're doing" at that particular moment. The information you can include in your tweet is virtually limitless, as long as you stick within the 140-character limitation that Twitter requires.

Get Personal

After you've set up your Twitter account, you'll first need to add your photo, a picture, or a logo to your profile. Next, create your bio, and be sure to make it interesting. Twitter gives you 140 characters to complete your profile, so be creative but still get to the point. It's one of the first things people look at when they're deciding whether or not to follow you.

Then, make sure to use the URL feature which allows your followers to easily access your company's Web site, your personal blogs, or whatever you decide to link it to. And whatever you do, don't forget to add some **PIZZAZZ** to your Twitter site. Under settings, you can customize your background or pick an option Twitter offers.

If your Twitter site is for personal use, think of Twitter as your very own personal Web site, where you can customize it to fit your personality, who you are, and make it your own.

If it's a corporate site, think of it as an extension of your corporate brand and identity. Either way, there are many ways to help make your Twitter site look and feel as you want the "Twitter Nation" to see you.

Contribute Value

It's great if you can contribute information to your community that provides value to your members, which will depend on who your followers are and what they expect to hear and see from you. Match your tweets to your audience. Use your smarts, your wit, and your passions to engage your followers. It's what tweeters want!

Who Tweets and How Often?

What does the typical tweeter look like? As of June 2009, according to Quantcast, Twitter reaches over 27 million people on a monthly basis. Within the U. S. here's what Twitter demographics look like:

- **74% of users are between the ages of 18 and 49 years old.**
- **Tweeters are 45% male, 55% female.**
- **78% are Caucasian.**
- **57% are either college graduates or post graduates.**
- **Tweeters average four tweets a day.**

Teens and Twitter

You'll find that most teens are very active on social networking sites such as MySpace and Facebook, and they love to text, but you won't find them using Twitter to that same degree.

According to CNBC.com, most under-twenties sign up for a Twitter account, but they don't actively send tweets because they realize no one is viewing their profile. Teens prefer the direct and private one-to-one communication of texting, Facebook, or MySpace as their social media platform.

Twitter Search

Resources are abundant when you're searching on Twitter, which is a great way for you to locate information quickly. Just type in your keyword in the Web or search box, and voila!—you'll instantly get more responses than you'll know what to do with. Here are a few things you can search for:

- **Interests**
- **Competitors**
- **Products or Services**
- **Your Favorite Star or News Commentator**
- **News or Information**
- **Trending Topics**
- **Help or Assistance**

Take a shot at trying an advanced search. It's easy, don't worry. Just for fun, see if anyone is tweeting about you, your company, or your product. While you're at it, why don't you search your key competitor? You could find some interesting intel, instantly elevating you to super-spy status . . . watch out James Bond! Too bad Q didn't have this cool tool to add to his repertoire of gadgets!

Trending Topics

Check out what's going on LIVE, right now under Twitters "Trending Topics." You'll be able to view real-time happenings at any moment, 24/7. Trending topics are the top ten topics of what the world of Twitter is discussing at that very moment in time. It's real-time 411 of what's hot and what's not!

What makes the trending topics list? If it's on the list, then the subject is within the top ten single words or phrases that are being tweeted the most at that particular time. Topics are always changing, depending on what's burning up the airwaves.

Purposefully tweeting within the trending topics can put you right into the stardom game, placing your updates for millions of viewers to see.

Following vs. Followers

One of the most important aspects of Twitter is to tweet helpful and important information worth reading. You'll want to attract relevant followers, so you have an audience to sing your tweets to. Potential followers (usually someone you request to follow first) will take a look at your picture and your profile to see if you are someone of interest and worth following. Make sure your profile is interesting and stands out and that your tweets are something worth reading.

As an example, if you're following a million people, and you only have ten followers, you could look as if you lack value or are not providing information worth receiving. You could also appear to be a spammer, which is exactly what you don't want. There are a few exceptions to this unspoken rule. People in the spotlight have millions of followers but are following only a few select people. Check out Oprah or the White House, and you'll see exactly what I mean.

Tweet Time

You'll want to post your tweets when traffic is the heaviest, so your voice is heard. According to Hubspot, Twitter users tend to pay most attention to Twitter traffic during the Eastern Time Zone's business hours Monday through Thursday, so jump on the Twitter "information highway" during those times.

Can't drive 65? Want to understand what Twitter is all about before you jump into heavy traffic? If this is the case for you, just take some time and jump onto Twitter during these busy times, watch what others are saying, and how they're delivering information. You'll be ready to hit the road in no time! Tweet Tweet!

Twittequette

Like any social network that's used in the public domain, there are unspoken rules of engagement you need to know **BEFORE** you start going "tweet crazy." Twitter also has its own set of unwritten rules and guidelines that dictates good and bad behavior, often referred to as twittequette.

Unlike texting, where you share your message person-to-person, your updates will be shared publicly for anyone to see, unless you are sending a direct message to someone who is following you. **IMPORTANT:** your tweets are permanent imprints that can't be edited once you've sent them. Take special notice that Twitter users are individuals and corporations, making it even more important to watch what you say at all times. Here's just a few of the dos and don'ts to think about:

Do . . .

- Have an interesting Twitter profile.
- Use Twitter for corporate and social networking as a main tool.
- Invite others to follow you, and follow whom you consider worth following.
- Share new, informative, or useful info.
- Build relationships and add value.

- Be entertaining, with class.

- Send direct messages for personal updates and personal information.

- Ask for help and give advice.

- Share interesting URLs, including a comment about what you're sharing.

- Direct others to a Web site or blog, if it's worth mentioning.

- Tweet your updates throughout the business week.

- Start or participate in a fund raiser.

- Tweet daily if you are providing valuable info such as dates, times, or locations.

- RT—When you find something great, share it by retweeting (RT) others' information, and be sure to acknowledge your source.

- Use Live Twitter to share what's happening at an event, while it's happening. (Don't forget to give credit when due.)

- Follow Twitter's official profile, then you'll always know what's going on.

Don't . . .

- Tweet if you need more than the alloted 140 characters.

- Give TMI—Don't tweet things you wouldn't want your mom to read; Twitter is a public medium.

- Over-tweet—Too many updates to Twitter become annoying and cause clutter.

- Be a Spammer—Report them to spam@twitter.com with details. Block or un-follow them too.

- Do a lot of boasting about you or your company. Too much is too much.

- Get too personal and over share; most people don't need to know those little details.

- Discuss politics—Keep your beliefs off of Twitter, unless politics is your passion or business.

- Dish about work, your boss, or your company (savvy recruiters keep an eye on most social networking sites when hiring).

- Personally tweet during work hours—Reserve updating for breaks and lunchtime.

- Tweet while driving or in social situations.

- Use a bunch of text lingo—Stick within the top fifty most-used abbreviations, and don't over-lingo within a single update.

Twitter and Your Business

Twitter can be an incredible tool to provide important information about your company, while at the same time revealing the human and real side of your organization.

Twitter is a great way to build relationships with those whom you might otherwise miss out on getting to know. Here are few things your company can do with Twitter:

- Provide customer support
- Network
- Distribute important warnings or service problems
- Provide important industry related info
- Refer your customers and others by retweeting their updates
- Add value to your service offerings
- Expand PR and marketing opportunities
- Share tips, promotions, and sales
- Make important announcements

Retweeting

Retweeting is exactly what it sounds like. You're reposting a tweet, which usually has a link attached, that was posted by someone else, so your followers can see it.

When you're retweeting, be sure to add a comment on why you're forwarding the message.

Important: Always be sure to give credit by adding "via" to the original source of the tweet your RTing.

Example: RT @original_authors_username: Check this out! I love this idea, it's fab: [then attach original message by way of a URL shortener].

Mobile Twitter

Once you've set up your new, no-cost Twitter account, completed your profile, and uploaded a picture, Twitter will give you the option to link your cell phone to your Twitter account. Once you do this, you'll be able to send and receive tweets from your cell phone. You'll also be able to select which tweets you want sent to your cell phone, so you don't get information overload. You'll also have the option to turn "delivery by text" off and on whenever you want.

The Twitter support site offers ample information. In order to text your tweets from your cell phone, add your cell phone number in your Twitter settings. Once you've added your number, you'll need to take the following steps, using Twitter's mobile text service:

1. Send the verification code in a text to the Twitter phone number assigned to you in your Twitter settings page.

2. Once you've verified your number with Twitter, you'll be able to send your tweets directly from your cell phone, and they'll

automatically post to your Twitter profile page and be distributed to your followers.

Note: You may use only one wireless phone number for each Twitter account.

If you're interested in using a mobile service with a little more oomph, use a dedicated mobile client specific to your cell phone:

BlackBerry: Check out Twitterberry online at www.orangatame.com or TinyTwitter at www.tinytwitter.com.

iPhone: Check out Tweetie at www.atebits.com or Twitterific at http://inconfactory.com

Palm Pre: For the Palm Pre, go to www.Palm.com for info and applications for use with Twitter.

Get Set Up

Twitter uses its own short and long codes (text-only numbers to send/receive Twitter updates), depending on where your wireless number is based. When you add your wireless number to Twitter, they will tell you which number to send your updates to:

- US – 40404
- Canada – 21212
- UK – Local Vodafone Users, 86444
- Other UK Services – +44 7624 1423

For other countries, contact Twitter for details.

If you have an international number, it's important to include "+" and your country code with your complete number, in order for your number to be recognized properly.

Example: +44 7624 801423

Note: When adding a country code to your number, in some countries, you must drop the leading zero.

Twitter doesn't charge for text messages, so your standard text-messaging rates will apply per the text plan you've selected with your wireless provider. Any applicable text charges will be billed from your provider, and the charge will depend on the text plan you have.

You'll see this disclaimer from Twitter:

> It's okay for Twitter to send txt messages to my phone. Standard rates apply.

What's the Difference Between a Tweet and a Text?

With Twitter, you can pay as much or as little attention to your tweets as you want without breaking any twittequette. However, if you're looking to attract followers, you're going to want to tweet your updates on a regular basis and keep them interesting and relative to what your followers expect to see from you. Because your tweets can be found by anyone on search, you'll be attracting new followers based on the appeal of your updates to other tweeters.

Even though you can easily text your tweets to your Twitter site, texts and tweets are used very differently, but they work beautifully together:

- ◎ Text messages are normally personal questions or comments privately sent to a single person.

- ◎ Tweets are typically posted information for many to read.

- ◎ When possible, texts are responded to immediately.

- ◎ Tweets are not always responded to and aren't as immediate as a TM.

- ◎ A text has 160 characters. If you go over, a second text is automatically sent.

- ◎ Twitter allows up to 140 characters to be sent in an update at one time.

- ◎ Twitter tends to attract bloggers and social media gurus.

- ◎ An average texter will send three hundred or more text messages a day, whereas an average tweeter sends one to four updates a day.

How Do I Find My Colleagues, Friends or Family?

When you create an account, you can search for users by name, or if you know it, by their Twitter username. At the top of your Twitter account page, click on the tab that says "Find People." At that point, you'll have two options:

1. **Find on Twitter**—Lets you search by name.

2. **Find on other networks**—This option checks your address book from Yahoo, Gmail, etc. against existing Twitter accounts.

Once you've found your friends, you can follow them, and hopefully they'll follow you back. If you can't find whom you're searching for, you can always click on the "invite by e-mail" tab and send a personal invitation that way. If you're still striking out, select the "suggested users" tab and see if anyone looks like a good fit.

Who Can Read My Updates?

Your updates are public, so anyone searching for anything on Twitter has the potential to read your tweets. You also have the option to protect your profile by approving followers and keeping your tweets private and out of search.

Can I Block People from Following Me?

You can easily block someone by selecting the person you want to block and clicking "block." When you block people, they won't be notified that you've blocked them, and they won't be able to send you messages or list you as someone they're following. However, if your account is public, they'll still be able to view your tweets.

Why Is There a Star at the End of Updates?

If you want to mark a tweet as a favorite, you can add a star. This is a great way to isolate all of your favorite twitter updates in one place, because when you click on the favorites link in your Twitter profile, they'll all show up there.

What Are @Replies?

When you want to direct your tweet to a specific person without having to go through the process of sending a direct message, send an @reply. Just understand that your tweet will still be public and searchable.

What Are Direct Messages?

Direct messages are private messages sent from one tweeter to another that you don't want the Twitter world to see. It's a way of messaging only one of your followers. Send a direct message when you want to tweet personal information like your phone number, your e-mail address, or your blog address.

D rbtlguide: Meet me at
Philippe at 33rd & E. 60th @ 9

For additional Twitter support, go to the Twitter Support site or the Twitter Blog at www.Twitter.com

Woo Hoo #9

hether you're interested in using Twitter to further your professional career, your company's presence in the Twitter world, or just looking to spread your wings and have some fun, Twitter can be the perfect information skyway for you.

For all you new "chicks" out there, find your flock to follow and join the wonderful world of Twitter.

Quick Tweet Lingo and Command Lesson

Nudge • Reminds a friend to update by asking them, on your behalf, what they're doing if they have not updated within twenty-four hours.

Tweet • An update or a single post.

Tweeters • People who post updates or tweets actively.

Retweet • To post someone else's tweet you like (RT).

Hashtag • The "#" symbol.

@username • Directs a Twitter at another person, and causes your Twitter to be saved in their replies tab.

D username • Sends a person a direct message.

WHOIS • Retrieves the profile info for any public user on Twitter.

FAV • Marks a person's last Twitter as a favorite. *

STATS • This command returns your number of Twitter followers, how many people you are following, and your bio info.

OH (overheard) • Use when tweeting a conversation you overheard.

INVITE • Sends a text message invite to a friend's cell phone. Example: Invite 415 555 1212

www.tinyurl.com • Allows you to enter a long URL to make it tiny, very helpful for Twitter updates.

www.bit.ly • Allows customizing shortened URLs and to keep track of click-throughs.

http://is.gd/ • Super shrinks URLs.

www.twitpic.com • Allows you to post photos through Twitter.

Txt un2 othrs as U would have
them txt un2 U

- Ten -

The Golden Rules of Texting

You're sitting at the dinner table with your family, and you notice your son has his cell phone on his lap; he's glancing down after every bite. Because "family dinner time" is one of those sacred, no-phone-allowed times, you politely ask him to stop. You get his famous eye roll in response to your request. The next thing you know, he's sneaking a peek at his phone. You quickly excuse yourself for a moment, go to the next room, grab your cell phone, and send him this text:

> Derek James, if u keep texting while u
> r at the dinner table, consider your
> cell phone mine 4 a week.

You then sit back at the dinner table, look at your son, and smile. You've just beat him at his own game! Nice job. You resolved the problem on Derek's level.

Manners Matter

Texting allows messages to be delivered quietly, no matter where you are. But text manners matter. Boost your text reputation by taking just a few minutes to read the basics of text etiquette:

- Never text at the dinner table or at any other family function.
- RSVP to your texts ASAP.
- Use :) & :(to avoid miscommunications.
- DON'T USE ALL CAPS UNLESS YOU'RE SHOUTING.
- Don't text @ weddings, funerals, school functions, or other text-free zones.
- Never text while dining out or in social situations. :(
- It's taboo to text while ordering @ Starbucks.
- Avoid the evil eye—don't text @ the theater in the middle of the movie.
- For relationship success—don't text on dates.
- Bring sexy back—DON'T text in the bedroom.

- Don't text in the boardroom—unless you're the boss.
- Get permission from your teacher to text @ school.

Tick Tock

A key to good texting is timeliness, so answer as quickly as possible. Replying to a text as soon as you can is expected in order to stay on the good side of your sender. When you're unable to answer a text immediately, you should respond once you become available. If you decide to wait and respond to your text a day or more later, you'll probably be sending a crystal clear message to your senders that their texts weren't important. If you become a text slacker, your friends and family will eventually stop attempting to contact you in this way.

Text-Free Zones

Show off your sweet side by not texting while engaged in conversations with anyone—it's just rude! Just as you shouldn't take any phone calls, you shouldn't be staring at your phone and texting, ignoring those you're with.

Avoid texting during meetings, school functions, and conferences. The only time using your BlackBerry, iPhone, or PDA is acceptable is if you're waiting for a critical, important text. Be sure your cell phone is on vibrate, and when your message arrives and you must respond immediately, excuse yourself for a few minutes and leave the room.

It goes without saying, but it needs to be said: don't even think about texting while in a place of worship, at weddings, or at funerals! These are serious text violations—to do so would be very disrespectful. Don't forget to change your cell phone to vibrate or silent mode to avoid that embarrassing "ding."

@ Restaurants

Sending text messages during a meal is telling your tablemates that they aren't as important or as interesting as the person on the other end of your cell phone. The people you're with will **NOT** appreciate that they're taking a backseat to an invisible third party. Remember, you're also being rude to the waiter who's trying to take your order. If you must answer a text, excuse yourself from the table to complete your connection.

There's nothing more annoying to the poor Starbucks barista who has to wait for you to finish your text while she's trying to take your order and make your drink, not to mention the people standing behind you waiting impatiently for their next caffeine fix. Go ahead and place your order first—you can always finish writing your text while your grande skinny hazelnut no-foam latte is being made. :)

@ The Movies or Theater

Movie and theater texting are big **NO-NOs**. While texts can seem completely private and unobtrusive, the light that goes on when you're in the dark comes through when a text is received, and the constant pecking at your cell phone during a movie is distracting to those around you. Just put your phone away. If you're expecting a must-read text like a dinner invite from that special someone, put the phone on silent, walk out to the lobby, and send your response:

> Im @ the movies, c u
> @ The Slanted Door @ 8.

Check the Time

When you want to send a late-night text, be aware that the ding and light may wake up your textee. It's always a good idea to be courteous and thoughtful when texting at night, so consider the time of your text and your audience. If you think it's possible someone might send you a text late at night when you'd rather remain fast asleep, you might want to put your phone on silent prior to hitting the pillow.

On a Date

You definitely shouldn't text while on a date, unless of course it's a blind date bomb, and you found your true love two tables over. You owe your SO (significant other) 100 percent of your attention, no exceptions. If you need to send a text, be respectful and wait until you go to the restroom to whip out your cell phone and start texting.

Family Time

Please don't text during dinnertime—the chef really won't appreciate it. Texting is also a no-no during important family discussions or family functions, and this goes for both parents and kids. Set boundaries and guidelines for text messaging in the home, and stick to them like glue.

@ School

The ability to use text messaging in the classroom varies from school to school, classroom to classroom, and teacher to teacher. Check your school's handbook or Web site for specific rules and regulations regarding the use of texting. It may be allowed between classes or on breaks but is usually not permissible during class time, unless you have that hip teacher who understands how to benefit from its use.

Students have been caught passing answers to test questions via text, so teachers are often unwilling to allow texting during classes without setting clear rules for their students, especially on exam days.

Special Occasions

Depending on the occasion, a text is always a nice little reminder to let that special someone know that you're thinking about him or her, especially if you haven't had the chance to call yet. For that extra special person or event, be sure to also send a greeting card or make a personal call in addition to your text—flowers are always nice too. :)

> Happy Birthday Cheryl! Let's Celebrate!
>
> Happy Anniversary Mom & Dad!
> I Love You Both!
>
> Great Job, Punk! I'm So Proud of You!
>
> Happy Birthday, Grandpa. I Love You

Text Tone

Like anything that's put in writing, text messages can easily be misunderstood or imply the wrong meaning, so be sure to give your text a quick read-through before you send it. Remember that text messaging is for quick and casual comments, not important conversations, so pick up the phone and make a call if it's more appropriate.

Don't use ALL CAPS unless you are truly mad, because sending a message in all caps is the same thing as SHOUTING! All caps are okay when you want to SHOUT or when you want to EMPHASIZE a word.

In the end, just use your best judgment when texting, and remember to add emoticons to clarify your emotions whenever appropriate. :)

Consideration Counts

Give some thought to the person to whom you're sending a text message. If the recipient isn't part of the Text Generation or is a "newbie" to text messaging, send simple messages that can be easily understood and limit the text lingo as much as possible.

I'm Sorry

Saying "I'm sorry" through a text message is probably not the best idea, unless you really can't reach that person any other way. Don't use texting as a substitute for personally sharing your regret for hurting someone's feelings or expressing your sympathy. Don't be wimpy; make the call or plan the visit. There's no substitute for an in-person conversation with someone you care about.

Lingo Overload

Too many text lingo abbreviations in one message can be annoying and not LOL funny: OMG, IDK, LOL :) TTYL :). Remember, a text is a substitute for a quick, brief, and informal note that usually requires an instant response.

When you're in the company of others, remember that there are very few things that absolutely can't wait. There are even fewer that are so immediate or important that every activity must be interrupted to read and respond to a text message.

Woo Hoo #10

Remember, a text is a substitute for a quick, brief, and informal note that usually requires an instant response.

When you're in the company of others, remember that there are very few things that absolutely can't wait. There are even fewer that are so immediate or important that every activity must be interrupted to read and respond to a text message.

Texting can be easily abused like anything else, so be aware of your surroundings and use your common sense to text tastefully. With a little effort, everyone around you will appreciate you even more than they already do!

Quick Text Lingo Lesson

BRB • Be Right Back

FWIW • For What It's Worth

GFTD • Gone For The Day

HB • Happy Birthday

KK • OK

LMK • Let Me Know

OMW • On My Way

PLZ • Please

SLAP • Sounds Like A Plan

TY • Thank You

WTH • What The Heck

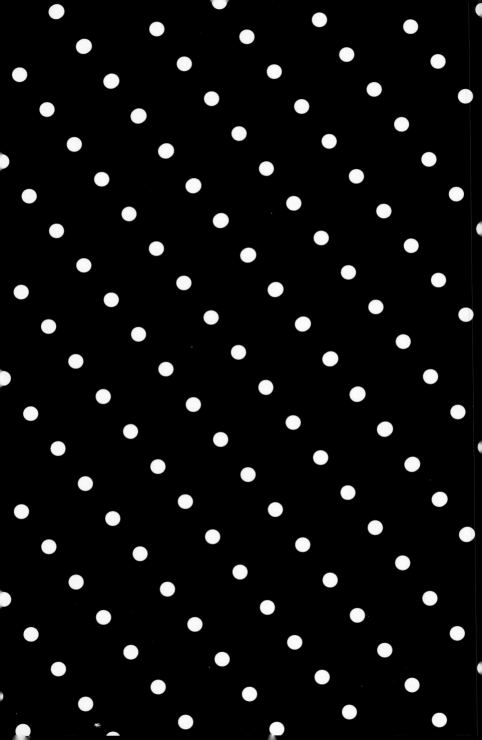

The List

Dictionaries for the New You

*C*onsider this your cheat sheet to unlock the mystery behind text lingo. Don't be offended at the common use of profanity in texting—you don't have to use it.

Maybe you want to become the next "high-wattage" text expert ever to walk the streets of your hometown. Or maybe you just want to learn the basics so you can send a simple message or translate a message you've received. Whatever your goal, you'll find more possibilities than you ever thought imaginable in "The List" sans the really offensive abbreviations.

The Top 50

- **2MOROW:**
  Tomorrow
- **2NITE:**
  Tonight
- **B4:**
  Before
- **B4N:**
  Bye For Now
- **B/C:**
  Because
- **BIZ:**
  Business
- **BRB:**
  Be Right Back
- **BTW:**
  By The Way
- **BTWN:**
  Between
- **BCNU:**
  Be Seeing You
- **BFF:**
  Best Friends Forever
- **BYAM:**
  Between You And Me

- **COT:**
  Circle Of Trust
- **CRS:**
  Can't Remember Sh**
- **C U LTR:**
  See You Later
- **DILLIGAS:**
  Do I Look Like I
  Give A Sh**
- **FWIW:**
  For What It's Worth
- **GTG:**
  Got To Go
- **GR8:**
  Great
- **HAH:**
  Big Laugh or Funny
- **IDK:**
  I Don't Know
- **IMO:**
  In My Opinion
- **JK:**
  Just Kidding
- **KK:**
  Ok, or I Agree

- **L8R:**
  Later

- **LMAO:**
  Laughing My Ass
  Off

- **LOL:**
  Laughing Out Loud

- **MINS:**
  Minutes

- **MYHOTY:**
  My Hat's Off To You

- **MYOB:**
  Mind Your Own
  Business

- **NP:**
  No Problem

- **OMG:**
  Oh My God

- **POV:**
  Point Of View

- **RBTL:**
  Read Between
  The Lines

- **ROFL:**
  Rolling On The
  Floor Laughing

- **SOL:**
  Sh** Out Of Luck

- **STBY:**
  Sucks To Be You

- **STRBKS/ *$s:**
  Starbucks

- **SWAK:**
  Sealed (or Sent) With
  A Kiss

- **TLC:**
  Tender Loving Care

- **TMI:**
  Too Much Information

- **TTFN:**
  Ta Ta For Now

- **TTYL:**
  Talk To You Later

- **THX, TX, THKS:**
  Thanks

- **TWTR:**
  Twitter

- **WTF:**
  What the F***

- **WTG:**
  Way To Go

- **WYWH:**
  Wish You Were Here

- **XOXO:**
  Hugs And Kisses

- **143:**
  I Love You

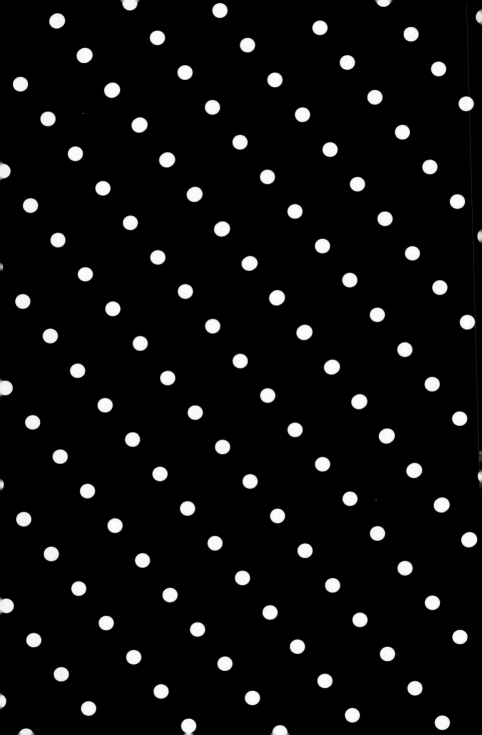

Text Lingo Dictionary

Numbers

- **02:** Two Cents' Worth
- **121:** One To One
- **1337:** Elite
- **10-4:** Agreed, Understood
- **143:** I Love You
- **14AA41:** One For All And All For One
- **182:** I Hate You
- **20:** Location
- **2b Or Not 2b:** To Be Or Not To Be
- **2GTBT:** Too Good To Be True
- **2MORO:** Tomorrow
- **2NITE:** Tonight
- **2U2:** To You Too
- **404:** No Clue, I Haven't A Clue
- **411:** Information
- **420:** Marijuana
- **459:** I Love You
- **4COL:** For Crying Out Loud
- **4EVA:** Forever
- **4EVER:** Forever
- **4NR:** Foreigner
- **5FS:** Five-Finger Salute
- **747:** Let's Fly

- **831:** I Love You (8 Letters, 3 Words, 1 Meaning)
- **86:** Thrown Out or Kicked Out
- **9:** Parent Is Watching
- **911:** Emergency
- **99:** Parent Is No Longer Watching
- **?:** Question or I Have A Question
- **@TEOTD:** At The End Of The Day

A

- **A/S/L/P:** Age/Sex/ Location/Picture
- **A3:** Any Place, Anywhere, Anytime
- **AAAAA:** American Association Against Acronym Abuse
- **AAF:** As A Friend
- **AAK:** Asleep At Keyboard
- **AAMOF:** As A Matter Of Fact
- **AAMOI:** As A Matter Of Interest
- **AAR8T:** At Any Rate
- **AAS:** Alive And Smiling

- **AATK:** Always At The Keyboard
- **AAYF:** As Always, Your Friend
- **AB:** Ass-Backwards
- **ABT2:** About To
- **ACD:** Alt Control Delete
- **ACE:** Access Control Entry
- **ACE:** Ace Insurance Company
- **ACH:** Automated Clearing House
- **ACK:** Acknowledgement
- **ACORN:** A Completely Obsessive, Really Nutty Person
- **ADAD:** Another Day, Another Dollar
- **ADIH:** Another Day In Hell
- **ADIP:** Another Day In Paradise
- **ADR:** Address
- **AEAP:** As Early As Possible
- **AFAGAY:** A Friend As Good As You
- **AFAHMASP:** A Fool And His Money Are Soon Parted
- **AFAIC:** As Far As I'm Concerned
- **AFAICS:** As Far As I Can See
- **AFAICT:** As Far As I Can Tell
- **AFAIK:** As Far As I Know
- **AFAIR:** As Far As I Remember
- **AFAIU:** As Far As I Understand
- **AFAP:** As Far As Possible
- **AFAYC:** As Far As You're Concerned
- **AFC:** Away From Computer
- **AFDN:** Any "F-ing" Day Now
- **AFGO:** Another "F-ing" Growth Opportunity

- **AFAIAA:** As Far As I Am Aware
- **AFIN:** A Friend In Need
- **AFJ:** April Fools' Joke
- **AFK:** Away From Keyboard
- **AFZ:** Acronym Free Zone
- **AGKWE:** And God Knows What Else
- **ALLRT:** All Right
- **AIH:** As It Happens
- **AMB:** As Mentioned Before
- **AIMP:** Always In My Prayers
- **AISB:** As I Said Before
- **AISE:** As I Said Earlier
- **AISI:** As I See It
- **AIR:** Adult In Room
- **AKA:** Also Known As
- **ALAP:** As Late As Possible
- **ALOL:** Actually Laughing Out Loud
- **ALOTBS:** Always Look On The Bright Side
- **ALTG:** Act Locally, Think Globally
- **AMAP:** As Many As Possible or As Much As Possible
- **AMBW:** All My Best Wishes
- **AMEXP:** American Express
- **AMF:** Adios Mother "F-er"
- **AML:** All My Love
- **AMZN:** Amazon.Com
- **AOAS:** All Of A Sudden
- **AOB:** Abuse Of Bandwidth
- **AP:** Affordable Pleasures or Apple Pie
- **APPS:** Applications
- **APPS:** Applications
- **ARC:** American Red Cross
- **AS:** Another Subject
- **ASAFP:** As Soon As "F-ing" Possible

- **ASAMOF:** As A Matter Of Fact
- **ASAP:** As Soon As Possible
- **ASAYGT:** As Soon As You Get This
- **ASL:** Age/Sex/Location
- **ATAB:** Ain't That A Bitch
- **ATC:** Any Two Cards
- **ATN:** All Together Now
- **ATSL:** Along The Same Line
- **ATSA:** At The Same Time
- **ATW:** All The Web or Around The Web
- **A2D:** Agree To Disagree
- **AWC:** After While, Crocodile
- **AWHFY:** Are We Having Fun Yet?
- **AWOL:** Absent Without Leave
- **AWTTW:** A Word To The Wise
- **AYC:** Aren't You Clever or Aren't You Cheeky
- **AYCE:** All You Can Eat
- **AYK:** As You Know
- **AYOR:** At Your Own Risk
- **AYSOS:** Are You Stupid or Something?
- **AYTMB:** And You're Telling Me This Because
- **AYV:** Are You Vertical?

- **B/C or BC:** Because
- **B&N:** Barnes & Noble
- **B&S:** Body & Soul
- **B4:** Before
- **B4N:** Bye For Now
- **B4U:** Before You
- **B4YKI:** Before You Know It

- **BAC:** Badass Chick
- **BAG:** Busting A Gut or Bigass Grin
- **BAK:** Back At Keyboard
- **BARB:** Buy Abroad But Rent In Britain
- **BAU:** Business As Usual
- **BB:** Be Back
- **BB4N:** Bye-Bye For Now
- **BBAMFIC:** Big, Badass Mother-"F-er" In Charge
- **BBB:** Bye-Bye Babe or Boring Beyond Belief
- **BBBG:** Bye-Bye, Be Good
- **BBFN:** Bye-Bye For Now
- **BBIAB:** Be Back In A Bit
- **BBIAF:** Be Back In A Few
- **BBIAS:** Be Back In A Sec
- **BBIAW:** Be Back In A While
- **BBL:** Be Back Later
- **BBR:** Burnt Beyond Repair
- **BBS:** Be Back Soon, or Bulletin Board Service
- **BBSD:** Be Back Soon, Darling
- **BBSL:** Be Back Sooner or Later
- **BBT:** Be Back Tomorrow
- **BBW:** Big, Beautiful Woman
- **BCBG:** Bon Chic, Bon Genre
- **BCBS:** Big Company, Big School
- **BCNU:** Be Seeing You
- **BCOZ:** Because
- **BD:** Big Deal or Baby Dance or Brain Drain
- **BDBI5M:** Busy Daydreaming, Back In 5 Minutes
- **BDC:** Big Dumb Company
- **BDN:** Big Damn Number
- **BEG:** Big Evil Grin
- **BF:** Boyfriend or Best Friend
- **BFD:** Big "F-ing" Deal

- **BFE:** Bum "F" Egypt
- **BFF:** Best Friends Forever
- **BFFN:** Best Friends For Now
- **BFFTTE:** Best Friends Forever Til The End
- **BFN:** Bye For Now
- **BHAG:** Big Hairy Audacious Goal
- **BHG:** Big-Hearted Guy or Big-Hearted Girl
- **BHIMBGO:** Bloody Hell, I Must Be Getting Old
- **BHFG:** Bald-Headed Fat Guy
- **BHOF:** Bald-Headed Old Fart
- **BI5:** Back In Five
- **BIBI:** Bye-Bye
- **BIBO:** Beer In, Beer Out
- **BIF:** Before I Forget
- **BIL:** Brother-In-Law
- **BION:** Believe It or Not
- **BIOYN:** Blow It Out Your Nose
- **BITCH:** Basically In The Clear, Homey
- **BITD:** Back In The Day
- **BITFO:** Bring It The "F" On
- **BKA:** Better Known As
- **BL:** Belly Laughing
- **BLKBRY:** BlackBerry
- **BM:** Bite Me
- **BMF:** Bad Mother "F-er"
- **BMGWL:** Busting My Gut With Laughter
- **BMOF:** Bite Me, Old Fart
- **BMOTA:** Bite Me On The Ass
- **BN:** Bare Necessities or Burn Notice
- **BNDN:** Been Nowhere, Done Nothing
- **BNF:** Big-Name Fan
- **BO:** Bugger Off or Body Odor

- **BOB:** Battery-Operated Boyfriend
- **BOB:** Body Off Baywatch
- **BOCTAAE:** But Of Course There Are Always Exceptions
- **BOFH:** Bastard Operator From Hell
- **BOHICA:** Bend Over Here It Comes Again
- **BION:** Believe It or Not
- **BOTEC:** Back Of The Envelope Calculation
- **BOTOH:** But On The Other Hand
- **BPLM:** Big Person, Little Mind
- **BR:** Bathroom
- **BRB:** Be Right Back
- **BRC:** Billy Ray Cyrus
- **BRT:** Be Right There
- **BS:** Big Smile or Bullsh** or Brain Strain
- **BSAAW:** Big Smile And A Wink
- **BSBD&NE:** Book Smart, Brain Dead & No Experience
- **BSEG:** Big Sh**-Eating Grin
- **BSF:** But Seriously, Folks
- **BSOD:** Blue Screen Of Death
- **BT:** Byte This or Bite This
- **BTA:** But Then Again or Before The Attacks
- **BTDT:** Been There, Done That
- **BTDTGTS:** Been There, Done That, Got The T-Shirt
- **BTHOOM:** Beats The Heck Out Of Me
- **BTN:** Better Than Nothing
- **BTOIYA:** Be There Or It's Your Ass

- **BTSOOM:** Beats The Sh** Out Of Me
- **BTTT:** Back To The Top
- **BTW:** By The Way
- **BTWBO:** Be There With Bells On
- **BTWITIAILW/U:** By The Way I Think I Am In Love With You
- **BW:** Best Wishes
- **BWDIK:** But What Do I Know
- **BWL:** Bursting With Laughter
- **BWO:** Black, White, or Other
- **BYKT:** But You Knew That
- **BYOA:** Bring Your Own Advil
- **BYOB:** Bring Your Own Beer or Bring Your Own Bottle
- **BYOW:** Build Your Own Web Site or Bring Your Own Wine
- **BZ:** Busy

C

- **C/S:** Change Of Subject
- **C4N:** Ciao For Now
- **CAAC:** Cool As A Cucumber
- **CALIGRL:** California Girl
- **CALYGY:** California Guy
- **CAS:** Crack A Smile
- **CB:** Chat Brat or Coffee Break
- **CBB:** Can't Be Bothered
- **CD9:** Code 9—It means Parents Are Around
- **CF:** Chick Flick/Coffee Freak
- **CFT:** Chick Flick Therapy
- **CFV:** Call For Vote
- **CHA:** Click Here, Asshole
- **CHG:** Change
- **CIAO:** Goodbye (in Italian)
- **CICO:** Coffee In, Coffee Out

- **CID:** Consider It Done or Crying In Disgrace
- **CIS:** Compuserve Info Service
- **CJ1:** Citation Jet 1
- **CJ2:** Citation Jet 2
- **CJ3:** Citation Jet 3
- **CJ4:** Citation Jet 4
- **CJX:** Citation Jet 10
- **CLM:** Career Limiting Move
- **CM:** Call Me
- **CMAP:** Cover My Ass, Partner
- **CMF:** Count My Fingers
- **CMIIW:** Correct Me If I'm Wrong
- **CMU:** Crack Me Up
- **CNP:** Continued In Next Post
- **COB:** Close Of Business
- **COD:** Change Of Direction
- **CONF #:** Confirmation Number
- **COS:** Church Of Scientology
- **COS:** Change Of Subject
- **COT:** Circle Of Trust
- **CRAFT:** Can't Remember A "F-ing" Thing
- **CRAT:** Can't Remember A Thing
- **CRB:** Come Right Back
- **CRBT:** Crying Really Big Tears
- **CRDCHK:** Credit Check
- **CRKBRY:** Crackberry
- **CRS:** Can't Remember Sh**
- **CS:** Career Suicide
- **CSA:** Cool, Sweet, Awesome
- **CSL:** Can't Stop Laughing
- **CSG:** Chuckle, Snicker, Grin
- **CSTCO:** Costco
- **CT:** Can't Talk
- **CTA:** Call To Action

- **CTC:** Choke The Chicken or Care To Chat
- **CTO:** Check This Out
- **CU:** See You or Cracking Up
- **CUATU:** See You Around The Universe
- **CUL8R:** See You Later
- **CULA:** See You Later, Alligator
- **CUOL:** See You Online
- **CUWTA:** Catch Up With The Acronyms
- **CUZ:** Because
- **CWOT:** Complete Waste Of Time
- **CWYL:** Chat With You Later
- **CX:** Cancelled
- **CY:** Calm Yourself
- **CYA:** Cover Your Ass or See Ya
- **CYE:** Check Your E-Mail
- **CYL:** See You Later
- **CYM:** Check Your Mail
- **CYO:** See You Online
- **CYT:** See You Tomorrow

D

- **DAMHIK:** Don't Ask Me How I Know
- **DARFC:** Ducking And Running For Cover
- **DBA:** Doing Business As
- **DBAB:** Don't Be A Bitch
- **DBD:** Don't Be Dumb
- **DBEYR:** Don't Believe Everything You Read
- **DC:** Damage Control
- **DD:** Due Diligence
- **DDD:** Direct Distance Dial
- **DDSOS:** Different Day, Same Old Sh**

- **DEF:** Definitely
- **DETI:** Don't Even Think It
- **DF:** Dear Friend
- **DGA:** Don't Go Anywhere
- **DGT:** Don't Go There
- **DGTG:** Don't Go There, Girlfriend
- **DGYF:** Damn Girl, You're Fine
- **DHYB:** Don't Hold Your Breath
- **DIAF:** Die In A Fire
- **DISAW:** Diane Sawyer
- **DIC:** Drunk In Charge
- **DIKU:** Do I Know You
- **DILLIGAD:** Do I Look Like I Give A Damn?
- **DILLIGAS:** Do I Look Like I Give A Sh**
- **DINK:** Double Incomes, No Kids
- **DIRFT:** Do It Right The First Time
- **DISTOL:** Did I Say That Out Loud?
- **DITR:** Dancing In The Rain
- **DITYID:** Did I Tell You I'm Distressed?
- **DIY:** Do It Yourself
- **DKDC:** Don't Know, Don't Care
- **DL:** Down Low (Bottom Line) or Download
- **DLTBBB:** Don't Let The Bed Bugs Bite
- **DLTM:** Don't Lie To Me
- **DMI:** Don't Mention It
- **DBL8:** Don't Be Late
- **DNC:** Does Not Compute
- **DND:** Do Not Disturb
- **DOC:** Drug Of Choice
- **DOE:** Depends On Experience

- **DOEI:** Goodbye (in Dutch)
- **DOR:** Department Of Redundancy
- **DP:** Domestic Partner
- **DPS:** Damage Per Second
- **DPYP:** Don't Poop Your Pants
- **DQMOT:** Don't Quote Me On This
- **DQYDJ:** Don't Quit Your Day Job
- **DR143:** Dr. Love
- **DRIB:** Don't Read If Busy
- **Dr. P:** Dr. Phil
- **DSTR8:** Damn Straight
- **DT:** Double Trouble
- **DTC:** Deep Throaty Chuckle
- **DTRT:** Do The Right Thing
- **DUI:** Driving Under The Influence
- **DUA:** Don't Use Acronyms
- **DURS:** Damn, You Are Sexy
- **DUSL:** Do You Scream Loud?
- **DUST:** Did You See That?
- **DWB:** Don't Write Back
- **DWBH:** Don't Worry, Be Happy
- **DWI:** Driving While Intoxicated
- **DWPKOTL:** Deep Wet Passionate Kiss On The Lips
- **DWS:** Driving While Stupid
- **DWWWI:** Surfing The World Wide Web While Intoxicated
- **DWYM:** Does What You Mean
- **DYFM:** Dude You Fascinate Me
- **DYHAB:** Do You Have A Boyfriend?
- **DYHAG:** Do You Have A Girlfriend?
- **DYJHIW:** Don't You Just Hate It When . . .
- **DYOFDW:** Do Your Own "F-ing" Dirty Work
- **DYSTSOTT:** Did You See The Size Of That Thing?

E

- **E123:** Easy As One, Two, Three
- **E2E:** Eye To Eye
- **ECOMP:** Electronic Workers Compensation
- **EE:** Electronic Emission
- **EFT:** Electronic Funds Transfer
- **EG:** Evil Grin
- **EIG:** Employers Insurance Group
- **EL:** Evil Laugh
- **EM:** Excuse Me
- **EMA:** E-Mail Address
- **EMBM:** Early Morning Business Meeting
- **EMFBI:** Excuse Me For Butting In
- **EMI:** Excuse My Ignorance
- **EML:** E-mail Me Later
- **EMSG:** E-Mail Message
- **EOD:** End Of Day
- **EOL:** End Of Life
- **ERLY:** Early
- **ESAD:** Eat Sh** And Die
- **ESEMED:** Every Second, Every Minute, Every Day
- **ESH:** Experience, Strength, And Hope
- **ESO:** Equipment Smarter Than Operator
- **ETA:** Estimated Time Of Arrival or Edited To Add

- **ETLA**: Extended Three-Letter Acronym
- **EVRY I**: Every One
- **EWI**: E-Mailing While Intoxicated
- **EZ**: Easy

- **F2F**: Face-To-Face
- **FAB**: Features, Attributes, Benefits
- **FAQ**: Frequently Asked Question
- **FASB**: Fast-Ass Son Of A Bitch
- **FAWC**: For Anyone Who Cares
- **FBI**: Federal Bureau Of Investigation or Female Body Inspector
- **FCFS**: First Come, First Served
- **FDGB**: Fall Down Go Boom
- **FE**: Fatal Error
- **FF**: Friends Forever
- **FF&PN**: Fresh Fields & Pastures New
- **FFS**: For "F" Sake
- **FIF**: "F" I'm Funny
- **FIFO**: First In, First Out
- **FIIK**: "F" If I Know
- **FIL**: Father-In-Law
- **FILTH**: Failed In London, Try Hong Kong
- **FIMH**: Forever In My Heart
- **FINE**: "F-ed" Up, Insecure, Neurotic, Emotional
- **FISH**: First In, Still Here
- **FITB**: Fill In The Blanks
- **FLA**: Four Letter Acronym
- **FML**: "F" My Life
- **FO**: "F" Off

- **FOAD**: "F" Off And Die
- **FOAF**: Friend Of A Friend
- **FOC**: Free Of Charge
- **FOFL**: Falling On Floor Laughing
- **FOMC**: Fell Off My Chair
- **FOCL**: Falling Off Chair Laughing
- **FORD**: Found On Road Dead or Fixed Or Repaired Daily
- **FOS**: Full Of Sh**
- **FS**: For Sale
- **FSBO**: For Sale By Owner
- **FSR**: For Some Reason
- **FTASB**: Faster Than A Speeding Bullet
- **FTBOMH**: From The Bottom Of My Heart
- **FTF**: Face To Face
- **FTFOI**: For The Fun Of It
- **FTL**: Faster Than Light
- **FTLOG**: For The Love Of God
- **FTN**: "F" That Noise
- **FTR**: For The Record
- **FTRF**: "F" That's Really Funny
- **FTTB**: For The Time Being
- **FTW**: For The Win
- **FUBAR**: "F-ed" Up Beyond All Recognition
- **FUBB**: "F-ed" Up Beyond Belief
- **FUD**: Fear, Uncertainty, & Disinformation
- **FUM**: "F-ed" Up Mess
- **FURTB**: Filled Up & Ready To Burst
- **FWB**: Friends With Benefits
- **FWD**: Forward
- **FWIW**: For What It's Worth

- **FYA:** For Your Amusement
- **FYE:** For Your Edification
- **FYF:** From Your Friend
- **FYI:** For Your Information
- **FYLTGE:** From Your Lips To God's Ears
- **FYM:** For Your Misinformation

- **G1:** Good One
- **G5:** Gulfstream V
- **GTG:** Got To Go
- **GTGLYS:** Got To Go, Love Ya So
- **G4I:** Go For It
- **G4N:** Good For Nothing
- **GA:** Go Ahead
- **GAB:** Got A Beer or Getting A Beer
- **GAL:** Get A Life
- **GALGAL:** Give A Little, Get A Little
- **GALHER:** Get A Load Of Her
- **GALHIM:** Get A Load Of Him
- **GANB:** Getting Another Beer
- **GAP:** Got A Picture
- **GAS:** Got A Second
- **GB:** Good Bridge
- **GBG:** Great Big Grin
- **GBH:** Great Big Hug
- **GBK:** Great Big Kiss
- **GC:** Good Crib
- **GDR:** Grinning, Ducking, & Running
- **GDRF:** Grinning, Ducking, & Running Fast
- **GDI:** God Damn It or God Damn Independent
- **GF:** Girlfriend

- **GFI:** Go For It
- **GFN:** Gone For Now
- **GFON:** Good For One Night
- **GFTD:** Gone For The Day
- **GFY:** Good For You
- **GG:** Gotta Go
- **GGA:** Good Game, All
- **GGL:** Google
- **GGN:** Gotta Go Now
- **GGOH:** Gotta Get Out Of Here
- **GGP:** Gotta Go Pee
- **GGRL:** Go Girl
- **GGRN:** Go Green
- **GI:** Google It
- **GIC:** Gift In Crib
- **GIDK:** Gee I Don't Know
- **GIGO:** Garbage In, Garbage Out
- **GJ:** Good Job or Great Job
- **GJP:** Good Job, Partner
- **GK:** Gayle King
- **GL:** Good Luck or Get Lost
- **GLA:** Good Luck, All
- **GLBT:** Gay, Lesbian, Bisexual, Transgender
- **GLG:** Good-Looking Girl
- **GLH:** Good-Looking & Handsome
- **GLGH:** Good Luck Good Hunting
- **GLYASDI:** God Loves You And So Do I
- **GM:** Good Morning or Good Move
- **GMA:** Good Morning America
- **GMAB:** Give Me A Break
- **GMAFB:** Give Me A "F-ing" Break

- **GMBA**: A Good Friend (In Italian)
- **GMTA**: Great Minds Think Alike
- **GMTFT**: Great Minds Think For Themselves
- **GOMB**: Get Off My Back
- **GN**: Good Night
- **GNBLFY**: Got Nothing But Love For You
- **GNOC**: Get Naked On Cam
- **GNSD**: Good Night, Sweet Dreams
- **GNST**: Good Night, Sleep Tight
- **GOI**: Get Over It
- **GOK**: God Only Knows
- **GOL**: Giggling Out Loud
- **GOS**: Gay Or Straight
- **GOWI**: Get On With It
- **GOYHH**: Get Off Your High Horse
- **GR8**: Great
- **GRAS**: Generally Recognized As Safe
- **GRATZ**: Congratulations
- **GRRR**: Growling
- **GSOAS**: Go Sit On A Snake
- **GSOH**: Good Sense Of Humor
- **GT**: Good Try
- **GTG**: Got To Go
- **GTGB**: Got To Go, Bye
- **GTGP**: Got To Go Pee
- **GTH**: Go To Hell
- **GTK**: Good To Know
- **GTM;** Giggle To Myself
- **GTP**: Get The Picture
- **GTRM**: Going To Read Mail
- **GTSY**: Glad To See You
- **GU**: Graphically Undesirable

- **GWI**: Get With It
- **GWS**: Get Well Soon
- **GYHOOYA**: Get Your Head Out Of Your Ass
- **GYPO**: Get Your Pants Off

- **H&K/HAK**: Hugs & Kisses
- **H/O**: Hold On
- **H/P**: Hold Please
- **H4U**: Hot For You
- **HAGD**: Have A Great Day
- **HAH**: Big Laugh, Funny
- **HAGN**: Have A Good Night
- **HAGO**: Have A Good One
- **HAND**: Have A Nice Day
- **HAR**: Hit And Run
- **HAWTLW**: Hello And Welcome To Last Week
- **HB**: Holy Batman or Happy Birthday or Hurry Back
- **HBASTD**: Hitting Bottom And Starting To Dig
- **HBB**: Hot Beyond Belief
- **HC**: Holy Crap
- **HD**: Hold
- **HF**: Have Faith, Hello Friend, or Have Fun
- **HHIS**: Hanging Head In Shame
- **HHOJ**: Ha-Ha, Only Joking
- **HHOK**: Ha-Ha, Only Kidding
- **HIG**: How's It Going or Hartford Insurance Group
- **HIH**: Hope It Helps
- **HIOOC**: Help, I'm Out Of Coffee
- **HITAKS**: Hang In There And Keep Smiling
- **HMFIC**: Head Mofo In Charge

- **HNTI**: How Nice This Is
- **HNTW**: How Nice That Was
- **HNY**: Happy New Year
- **HO**: Hang On or Hold On or Hooker
- **HOHA**: Hollywood Hacker
- **HOIC**: Hold On, I'm Coming
- **HOT PIC**: Hot, Sexy, or Naked Picture
- **HOYEW**: Hanging On Your Every Word
- **HP**: Higher Power
- **HPPO**: Highest Paid Person In Office
- **HSIK**: How Should I Know?
- **HT**: Hi There
- **HTB**: Hang The Bastards
- **HTF**: Hartford Insurance Company
- **HTH**: Hope This (or That) Helps
- **HTNOTH**: Hit The Nail On The Head
- **HTY**: Hottie or A Cute Person
- **HUA**: Heads Up, Ace or Head Up Ass
- **HUGZ**: Hugs
- **HUYA**: Head Up Your Ass
- **HWG**: Here We Go
- **HWGA**: Here We Go Again

- **I&I**: Intercourse & Inebriation
- **IDL**: Ideal
- **IAC**: In Any Case or I Am Confused
- **IAE**: In Any Event
- **IAITS**: It's All In The Subject

- **IANAC**: I Am Not A Crook
- **IANAE**: I Am Not An Expert
- **IANAL**: I Am Not A Lawyer
- **IASAP4U**: I Always Say A Prayer For You
- **IAT**: I Am Tired
- **IAW**: I Agree With or In Accordance With
- **IAWW**: In A Woman's World
- **IAYM**: I Am Your Master
- **IBGYBG**: I'll Be Gone, You'll Be Gone
- **IBIWISI**: I'll Believe It When I See It
- **IBRB**: I'll Be Right Back
- **IBT**: In Between Technology
- **IBTC**: Itty Bitty Titty Committee
- **IBTS**: In Between The Sheets
- **IBTD**: I Beg To Differ
- **IBTL**: In Before The Lock
- **IC**: In Character or I See
- **ICBW**: I Could Be Wrong
- **ICUB**: I'll Call You Back
- **ICYC**: In Case You're Curious or In Case You Care
- **ID10T**: Idiot
- **IDC**: I Don't Care
- **IDGARA**: I Don't Give A Rat's Ass
- **IDGI**: I Don't Get It or I Don't Get Involved
- **IDK**: I Don't Know
- **IDKY**: I Don't Know You
- **IDM**: It Does Not Matter
- **IDST**: I Didn't Say That
- **IDTA**: I Did That Already
- **IDTS**: I Don't Think So

- **IF/IB:** In The Front or In The Back
- **IFAB:** I Found A Bug
- **IFON:** iPhone
- **IGGP:** I Gotta Go Pee
- **IGTP:** I Get The Point
- **IHA:** I Hate Acronyms
- **IHNO:** I Have No Opinion
- **IHU:** I Hear You
- **IIABDFI:** If It Ain't Broke, Don't Fix It
- **IIIO:** Intel Inside, Idiot Outside
- **IIMAD:** If It Makes Any Difference
- **IIR:** If I Remember or If I Recall
- **IIRC:** If I Remember Correctly or If I Recall Correctly
- **IIT:** Is It Tight
- **IITB:** It's In The Bank
- **IITLYTO:** If It's Too Loud, You're Too Old
- **IITYWIMWYBMAD:** If I Tell You What It Means, Will You Buy Me A Drink?
- **IITYWYBMAD:** If I Tell You, Will You Buy Me A Drink?
- **IIWM:** If It Were Me
- **IJPMP:** I Just Pissed My Pants
- **IJWTK:** I Just Want To Know
- **IJWTS:** I Just Want To Say
- **IKALOPLT:** I Know A Lot Of People Like That
- **IKWYM:** I Know What You Mean
- **IKYABWAI:** I Know You Are But What Am I?
- **ILA:** I Love Acronyms
- **ILF/MD:** I Love Female/Male Dominance
- **ILU:** I Love You
- **ILUAAF:** I Love You As A Friend
- **ILY:** I Love You
- **IM:** Instant Messaging or Immediate Message
- **IM2BZ2P:** I Am Too Busy To Pee
- **IMA:** I Might Add
- **IMAO:** In My Arrogant Opinion
- **IMCO:** In My Considered Opinion
- **IME:** In My Experience
- **IMEZRU:** I'm Easy, Are You?
- **IMHEIUO:** In My High, Exalted, Informed, Unassailable Opinion
- **IMHO:** In My Humble Opinion
- **IMNERHO:** In My Never-Even-Remotely Humble Opinion
- **IMNSHO:** In My Not-So-Humble Opinion
- **IMO:** In My Opinion
- **IMOO:** In My Own Opinion
- **IMPOV:** In My Point Of View
- **IMRU:** I Am, Are You?
- **IMS:** I'm Sorry
- **INBD:** It's No Big Deal
- **INMP:** It's Not My Problem
- **INNW:** If Not Now, When
- **INPO:** In No Particular Order
- **INUCOSM:** It's No Use Crying Over Spilt Milk

- **IOH**: I'm Outta Here
- **ION**: Index Of Names
- **IOUD**: Inside, Outside, Upside Down
- **IOW**: In Other Words
- **IRL**: In Real Life
- **ISAGN**: I See A Great Need
- **ISH**: Insert Sarcasm Here
- **ISO**: In Search Of
- **ISS**: I Said So or I'm So Sure
- **ISTM**: It Seems To Me
- **ISTR**: I Seem To Remember
- **ISWYM**: I See What You Mean
- **ISYALS**: I'll Send You A Letter Soon
- **ITA**: I Totally Agree
- **ITFA**: In The Final Analysis
- **ITIGBS**: I Think I'm Going To Be Sick
- **ITK**: In The Know
- **ITM**: In The Money
- **ITS**: Intense Text Sex
- **IT'S A D8**: It's A Date
- **TSFWI**: If The Shoe Fits, Wear It
- **ITUB**: I'll Text You Back
- **IUM**: If You Must
- **IWALU**: I Will Always Love You
- **IWBNI**: It Would Be Nice If
- **IWIWU**: I Wish I Was You
- **IWSN**: I Want Sex Now
- **IYD**: In Your Dreams
- **IYKWIM**: If You Know What I Mean
- **IYKWIMAITYD**: If You Know What I Mean, And I Think You Do
- **IYO**: In Your Opinion
- **IYSS**: If You Say So

- **J/C**: Just Checking
- **J/J**: Just Joking
- **J/K or JK**: Just Kidding
- **J/P**: Just Playing
- **J/W**: Just Wondering
- **J2LYK**: Just To Let You Know
- **J4F**: Just For Fun
- **J4G**: Just For Grins
- **J4T or JFT**: Just For Today
- **J5M**: Just Five Minutes
- **JAD**: Just Another Day
- **JAM**: Just A Minute
- **JAS**: Just A Second
- **JC**: Just Curious or Just Chilling or Jesus Christ
- **JDI**: Just Do It
- **JTLS**: Jetless
- **JFI**: Just For Information
- **JIC**: Just In Case
- **JK**: Just Kidding
- **JM2C**: Just My 2 Cents
- **JMBA**: Jamba Juice
- **JMO**: Just My Opinion
- **JOOTT**: Just One Of Those Things
- **JP**: Just Playing
- **JSU**: Just Shut Up
- **JSYK**: Just So You Know
- **JT**: Just Teasing
- **JTLYK**: Just To Let You Know
- **JTOL**: Just Thinking Out Loud
- **JTOU**: Just Thinking Of You
- **JW**: Just Wondering

- **K**: Ok
- **KB**: Kick Butt or Kick Boxing

- **KEWL:** Cool
- **KFY or K4Y:** Kiss For You
- **KHYF:** Know How You Feel
- **KIA:** Killed In Action
- **KIR:** Keep It Real
- **KISS:** Keep It Simple Stupid
- **KIT:** Keep In Touch
- **KK:** OK or Kiss Kiss
- **KMA:** Kiss My Ass
- **KMP:** Keep Me Posted
- **KMUF:** Kiss Me, You Fool
- **KOTC:** Kiss On The Cheek
- **KOTL:** Kiss On The Lips
- **KPC:** Keeping Parents Clueless
- **KUTGW:** Keep Up The Good Work
- **KWIM:** Know What I Mean?
- **KWSTA:** Kiss With Serious Tongue Action
- **KYFC:** Keep Your Fingers Crossed
- **KYPO:** Keep Your Pants On

- **L:** Laugh
- **L8R:** Later
- **LABATYD:** Life's A Bitch And Then You Die
- **LAQ:** Lame-Ass Quote
- **LBR:** Little Boys' Room
- **LD:** Long Distance or Later Dude
- **LDR:** Long-Distance Relationship
- **LF:** Let's "F"
- **LFTI:** Looking Forward To It
- **LGMAS:** Lord, Give Me A Sign
- **LGR:** Little Girls Room

- **LHO:** Laughing Head Off
- **LHOS:** Lets Have Online Sex
- **LHSO:** Let's Have Sex Online
- **LIFO:** Last In, First Out
- **LIFW:** Lunch Is For Wimps
- **LIS:** Laughing In Silence
- **LJBF:** Let's Just Be Friends
- **L/M:** Left Message
- **LMAO:** Laughing My Ass Off
- **LMFAO:** Laughing My "F-ing" Ass Off
- **LMHO:** Laughing My Head Off
- **LMIRL:** Let's Meet In Real Life
- **LMK:** Let Me Know
- **LMSO:** Laughing My Socks Off
- **LNBM:** Late Night Business Meeting
- **LOL:** Laughing Out Loud or Lots Of Love
- **LOLA:** Laugh Out Loud Again
- **LOML:** Love Of My Life
- **LONH:** Lights On, Nobody Home
- **LOOL:** Laughing Outrageously Out Loud
- **LOPSOD:** Long On Promises, Short On Delivery
- **LORE:** Learn Once, Repeat Everywhere
- **LOU:** Laughing Over You
- **LPOS:** Lazy Piece Of Sh**
- **LSHMBH:** Laughing So Hard My Belly Hurts
- **LSV:** Language, Sex, Violence
- **LTIC:** Laughing Till I Cry
- **LTM:** Laughing To Myself
- **LTNS:** Long Time No See

- **LTR:** Long-Term Relationship
- **LTS:** Laughing To Self
- **LTTIC:** Look, The Teacher Is Coming
- **LU:** Listen Up
- **LUSM:** Love You So Much
- **LWR:** Launch When Ready
- **LY:** Love You
- **LY4E:** Love You Forever
- **LYA:** Love You All
- **LYB:** Love You, Babe
- **LYBL:** Live Your Best Life
- **LYCYLBB:** Love You, See You Later, Bye-Bye
- **LYL:** Love You Lots
- **LYLAB:** Love You Like A Brother
- **LYLAS:** Love You Like A Sister
- **LYLB:** Love You Later, Bye
- **LYMI:** Love You, Mean It
- **LYWAMH:** Love You With All My Heart

- **M2NY:** Me Too, Not Yet
- **M4C:** Meet For Coffee
- **M8 or M8S:** Mate or Mates
- **MA:** Mature Audience
- **MAYA:** Most Advanced, Yet Accessible
- **MB:** Message Board
- **MBN:** Must Be Nice
- **MBNZ:** Mercedes Benz
- **MBRFN:** Must Be Real "F-ing" Nice
- **MC:** Miley Cyrus
- **MEGO:** My Eyes Glaze Over
- **MFD:** Multifunction Device

- **MFWIC:** Mo Fo Who's In Charge
- **MHBFY:** My Heart Bleeds For You
- **MHOTY:** My Hat's Off To You
- **MIA:** Missing In Action
- **MIHAP:** May I Have Your Attention Please?
- **MIL:** Mother-In-Law
- **MILF:** Mom I'd Like To "F"
- **MINS:** Minutes
- **MIRL:** Meet In Real Life
- **MITIN:** More Info Than I Needed
- **MKOP:** My Kind Of Place
- **MLA:** Multiple Letter Acronym
- **MLAS:** My Lips Are Sealed
- **MM:** Market Maker
- **MO:** Move On
- **MOF:** Matter Of Fact
- **MOOS:** Member Of The Opposite Sex
- **MORF:** Male or Female
- **MOS:** Mom Over Shoulder
- **MOSS:** Member(s) Of The Same Sex
- **MOTD:** Message Of The Day
- **MOTOS:** Member(S) Of The Opposite Sex
- **MOTSS:** Member(S) Of The Same Sex
- **MRA:** Moving Right Along
- **MSFT:** Microsoft
- **MSG:** Message
- **MSMD:** Monkey See, Monkey Do
- **MSNUW:** Mini-Skirt No Underwear

- **MTAPF:** More Than A Pretty Face
- **MTBF:** Mean Time Before Failure
- **MTF:** More To Follow
- **MTFBWY:** May The Force Be With You
- **MTLA:** My True Love Always
- **MUBR:** Messed Up Beyond Recognition
- **MUSM:** Miss You So Much
- **MWBRL:** More Will Be Revealed Later
- **MYL:** Mind Your Language
- **MYOB:** Mind Your Own Business

N

- **N/A:** Not Applicable or Not Affiliated
- **N/M or NM:** Nothing Much
- **N/T:** No Text
- **N1:** Nice One
- **N2M:** Not To Mention or Not Too Much
- **NAB:** Not A Blonde
- **NADT:** Not A Damn Thing
- **NALOPKT:** Not A Lot Of People Know That
- **NATCH:** Naturally
- **NAVY:** Never Again Volunteer Yourself
- **NAZ:** Name, Address, Zip or Nasdaq
- **NBD:** No Big Deal
- **NBFAB:** Not Bad For A Beginner
- **NBIF:** No Basis In Fact

- **NBLFY:** Nothing But Love For You
- **NC:** Nice Crib
- **NCG:** New College Graduate
- **NCP:** Nincompoop
- **NDN:** Indian
- **NEWY:** AnyWay
- **NE1:** Anyone
- **NE14KFC:** Anyone For KFC?
- **NE2HV:** Need To Have
- **NESEC:** Any Second
- **NEWS:** North, East, West, South
- **NFW:** No Feasible Way
- **NH:** Nice Hand
- **NHOH:** Never Heard Of Him/Her
- **NIFOC:** Nude In Front Of The Computer
- **NIGY:** Now I've Got You
- **NIH:** Not Invented Here
- **NIM:** No Internal Message
- **NIMBY:** Not In My Backyard
- **NIMJD:** Not In My Job Description
- **NIMQ:** Not In My Queue
- **NIMY:** Never In A Million Years
- **NISM:** Need I Say More
- **NITL:** Not In This Lifetime
- **NIYWD:** Not In Your Wildest Dreams
- **NLL:** Nice Little Lady
- **NM:** Never Mind or Nothing Much
- **NME:** Enemy
- **NMH:** Not Much Here
- **NMHJC:** Not Much Here, Just Chilling
- **NMP:** Not My Problem

- **NMTE**: Now More Than Ever
- **NMU?**: Not Much, You?
- **NN**: Not Now
- **NNWW**: Nudge, Nudge, Wink, Wink
- **NOFI**: No Offence Intended
- **NOOB**: Newbie, New Person
- **NOYB**: None Of Your Business
- **NP**: No Problem
- **NQA**: No Questions Asked
- **NR**: Nice Roll or Nice Rig
- **NRG**: Energy
- **NRN**: No Reply Necessary
- **NS**: Nice Set
- **NSA**: No Strings Attached
- **NSS**: No Sh**, Sherlock
- **NSTLC**: Need Some Tender Loving Care
- **NTA**: Not This Again
- **NTIM**: Not That It Matters
- **NTIMM**: Not That It Matters Much
- **NTK**: Nice To Know
- **NTM**: Not That Much
- **NTTAWWT**: Not That There's Anything Wrong With That
- **NTW or N2W**: Not To Worry
- **NTYMI**: Now That You Mention It
- **NUB**: New Person To A Site or Game
- **NUFF**: Enough Said
- **NVM**: Never Mind
- **NVNG**: Nothing Ventured, Nothing Gained
- **NW**: No Way
- **NWAL**: Nerd Without A Life

- **NWR**: Not Work Related
- **NYC**: Not Your Concern or New York City

- **O**: Oh or Opponent or Over
- **OAO**: Over And Out
- **OAUS**: On An Unrelated Subject
- **OB**: Obligatory
- **OBE**: Overcome By Events
- **OBO**: Or Best Offer
- **OBTW**: Oh, By The Way
- **OCD**: Obsessive Compulsive Disorder
- **ODTAA**: One Damn Thing After Another
- **OIC**: Oh, I See
- **OK**: Ok or All Correct
- **OL**: Old Lady
- **OLL**: Online Love
- **OM**: Old Man
- **OMB**: Oh My Buddha
- **OMDB**: Over My Dead Body
- **OMG**: Oh My God
- **OMIF**: Open Mouth, Insert Foot
- **OMIK**: Open Mouth, Insert Keyboard
- **OML**: Oh My Lord
- **OMW**: On My Way
- **ONID**: Oh No I Didn't
- **ONNA**: Oh No, Not Again
- **ONNTA**: Oh No, Not This Again
- **ONUD**: Oh No You Didn't
- **OO**: Over & Out or Over & Over
- **OOAK**: One Of A Kind

- **OOC**: Out Of Control or Out Of Character
- **OOI**: Out Of Interest
- **OOO**: Out Of Office
- **OOS**: Out Of Stock
- **OOTB**: Out Of The Box or Out Of The Blue
- **OSIF**: Oh Sh**, I Forgot
- **OSINTOT**: Oh Sh**, I Never Thought Of That
- **OST**: On Second Thought
- **OT**: Off Topic
- **OTC**: Over The Counter
- **OTE**: On The Edge
- **OTF**: Off The Floor
- **OTH**: Off The Hook
- **OTL**: Out To Lunch
- **OOT**: Out Of Touch
- **OPRH**: Oprah Winfrey
- **OTP**: On The Phone
- **OTOH**: On The Other Hand
- **OTR**: On The Radar
- **OTT**: Over The Top
- **OTTOMH**: Off The Top Of My Head
- **OTW**: Off The Wall
- **OW**: Oprah Winfrey
- **OUSU**: Oh, You Shut Up
- **OZ**: Australia or Ozzy Ozbourne

- **P**: Partner or Pee
- **P&C**: Private & Confidential
- **P-ZA**: Pizza
- **P911**: Parent Alert
- **PA**: Parent Alert
- **PAL**: Parents Are Listening

- **PANS**: Pretty Awesome New Stuff
- **PAW**: Parents Are Watching
- **PB**: Potty Break
- **PBB**: Parent Behind Back
- **PBJ**: Peanut Butter & Jelly or Pretty Boy Jock
- **PC**: Personal Computer or Politically Correct
- **PD**: Purpose-Driven or Public Domain
- **PDOMA**: Pulled Directly Out Of My Ass
- **PDQ**: Pretty Darn Quick
- **PDS**: Please Don't Shout
- **PEEPS**: People
- **PERF**: Perfect
- **PFA**: Pulled From Ass or Please Find Attached
- **PHAT**: Pretty Hot And Tempting
- **PHLTN**: Paris Hilton
- **PIA**: Pain In The Ass
- **PIF**: Paid In Full
- **PIMP**: Peeing In My Pants
- **PIMPL**: Peeing In My Pants Laughing
- **PIN**: Person In Need
- **PIR**: Parent In Room
- **PITA**: Pain In The Ass
- **PL**: Power Lunch
- **PLO**: Peace, Love, Out
- **PLOS**: Parents Looking Over Shoulder
- **PLS or PLZ**: Please
- **PM**: Personal Message or Private Message
- **PMBI**: Pardon My Butting In
- **PMF**: Pardon My French or Pure, Freaking Magic

- **PMFJI**: Pardon Me For Jumping In
- **PMJI**: Pardon My Jumping In
- **PML**: Pissing Myself Laughing
- **PMP**: Peeing My Pants
- **PMSL**: Pissed Myself Laughing
- **PNHD**: Pin Head
- **PO**: Piss Off
- **POAHF**: Put On A Happy Face
- **POMS**: Parent Over My Shoulder
- **PONA**: Person Of No Account
- **POOF**: Good-Bye or Gone
- **POS**: Parent Over Shoulder or Piece Of Sh**
- **POTUS**: President Of The United States
- **POV**: Point Of View
- **PPL**: People
- **PREZ**: President
- **PRW**: Parents Are Watching
- **PS**: Post Script
- **PSA**: Public Service Announcement
- **PTH**: Prime Tanning Hours
- **PTMM**: Please Tell Me More
- **PTP**: Pardon The Pun
- **P/U**: Pick Up
- **PU**: Stinky or That Stinks!
- **PWR**: Power
- **PYT**: Pretty Young Thing

Q

- **Q**: Question
- **QL**: Quit Laughing
- **QOTD**: Quote Of The Day
- **QQ**: Quick Question
- **QT**: Cutie
- **QYB**: Quit Your Bitching

R

- **R U THERE?**: Are You There?
- **R&D**: Research & Development
- **R&R**: Rest & Relaxation
- **RB**: Red Bull
- **RBAY**: Right Back At You
- **RBTL**: Read Between The Lines
- **RC**: Remote Control
- **RE**: Regarding
- **RESO**: Reservation
- **RFD**: Request For Discussion
- **RFTS**: Reach For The Stars
- **RGR**: Roger, Sounds Good
- **RHIP**: Rank Has Its Privileges
- **RTZ CRLTN**: The Ritz-Carlton
- **RIYL**: Recommended If You Like
- **R2BA**: Right To Bear Arms
- **RL**: Real Life
- **RLF**: Real Life Friend
- **RM**: Remake
- **RMLB**: Read My Lips, Baby
- **RMMA**: Reading My Mind Again
- **RMMM**: Read My Mail, Man
- **RN**: Right Now
- **RNN**: Reply Not Necessary
- **ROBROB**: Robin Roberts
- **ROFL**: Rolling On Floor Laughing or Rolling On The Floor Laughing
- **ROFLMAO**: Rolling On The Floor Laughing My Ass Off
- **ROFLMFAO**: Rolling On The Floor Laughing My "F-ing" Ass Off

- **ROFLOL:** Rolling On The Floor Laughing Out Loud
- **ROGL:** Rolling On The Ground Laughing
- **ROGLMAO:** Rolling On The Ground Laughing My Ass Off
- **ROTM:** Right On The Money
- **RR:** Range Rover
- **RRQ or RRR:** Return Receipt Request
- **RSN:** Real Soon Now
- **RSVP:** *Répondez S'il Vous Plaît*
- **RT:** Real Time
- **RTBM:** Read The Bloody Manual
- **RTBS:** Reason To Be Single
- **RTFAQ:** Read The FAQ
- **RTFM:** Read The "F-ing" Manual
- **RTFQ:** Read The "F-ing" Question
- **RTH:** Release The Hounds
- **RTM:** Read The Manual
- **RU:** Are You
- **RU/18:** Are You Over Eighteen?
- **RUH:** Are You Horny?
- **RUMORF:** Are You Male Or Female?
- **RUNTS:** Are You Nuts?
- **RUOK:** Are You OK?
- **RUS:** Are You Serious?
- **RUT:** Are You There?
- **RUUP4IT:** Are You Up For It?
- **RWI:** Roll With It
- **RX:** Drugs or Prescription
- **RYM:** Read Your Manual

- **RYO:** Roll Your Own
- **RYS:** Read Your Screen

S

- **S:** Smile
- **S2R:** Send To Receive
- **S2U:** Same To You
- **S&:R** Send & Receive
- **SAHM:** Stay At Home Mom
- **SAIA:** Stupid Ass In Action
- **SAPFU:** Surpassing All Previous Foul-Ups
- **SB:** Stand By
- **SBI:** Sorry 'Bout It
- **SBT:** Sorry 'Bout That
- **SCNR:** Sorry, Couldn't Resist
- **SEC:** Wait A Second, Give Me A Second
- **SED:** Said Enough, Darling
- **SEG:** Sh** Eating Grin
- **SEP:** Somebody Else's Problem
- **SETE:** Smiling Ear To Ear
- **SEWAG:** Scientifically Engineered, Wild-Ass Guess
- **SF:** Surfer Friendly or Science Fiction
- **SFAIAA:** So Far As I Am Aware
- **SFETE:** Smiling From Ear To Ear
- **SFTS:** Shoot For The Stars
- **SFX:** Sound Effects
- **SG:** Solid Ground
- **SH:** Sh** Happens
- **SHH:** Be Quiet
- **SHB:** Should Have Been
- **SHID:** Slap Head In Disgust
- **SHMILY:** See How Much I Love You
- **SHTY:** Shorty

- **SIC:** Spelling Is Correct
- **SICL:** Sitting In Chair Laughing
- **SICS:** Sitting In Chair Snickering
- **SII:** Seriously Impaired Imagination
- **SIL:** Sister-In-Law
- **SIP:** Skiing In Powder
- **SIT:** Stay In Touch
- **SITCOMS:** Single Income, Two Children, OppressIve Mortgage
- **SITD:** Still In The Dark
- **SIUYA:** Shove It Up Your Ass
- **SLKR:** Slacker
- **SK8R:** Skater
- **SL:** Second Life
- **SLAP:** Sounds Like A Plan
- **SLAW:** Sounds Like A Winner
- **SLRK:** Smart Little Rich Kid
- **SLOM:** Sticking Leeches On Myself
- **SLT:** Something Like That
- **SM:** Senior Moment
- **SMB:** Suck My Balls
- **SME:** Subject Matter Expert
- **SMH:** Shaking My Head
- **SMOP:** Small Matter Of Programming
- **SNAG:** Sensitive New-Age Guy
- **SNERT:** Snotty-Nosed, Egotistical, Rotten Teenager
- **SO:** Significant Other (Spouse, Boy/Girlfriend)
- **SOB:** Son Of A B*tch
- **SOBT:** Stressed Out Big-Time
- **SOGOP:** Sh** Or Get Off The Pot
- **SOH:** Sense Of Humor

- **SOHF:** Sense Of Humor Failure
- **SOI:** Self-Owning Idiot
- **SOIAR:** Sit On It And Rotate
- **SOL:** Sh** Out Of Luck
- **SOMY:** Sick Of Me Yet?
- **SOOYA:** Snake Out Of Your Ass
- **SOP:** Standard Operating Procedure
- **SORG:** Straight Or Gay?
- **SOS:** Same Old Sh** or 911
- **SOT:** Short On Time
- **SOTMG:** Short On Time, Must Go
- **SOW:** Speaking Of Which or Statement Of Work
- **SOZ:** Sorry
- **SP:** Stimulus Package
- **SPI:** Self-Proclaimed Idiot
- **SRO:** Standing Room Only
- **SRSLY:** Seriously
- **SSC:** Super Sexy Cute
- **SSDD:** Same Sh**, Different Day
- **SSIA:** Subject Says It All
- **ST:** Straight Talk
- **STATS:** Statistics
- **STBY:** Sucks To Be You
- **STD:** Seal The Deal
- **STFU:** Shut The "F" Up
- **STM:** Spank The Monkey
- **STPPYNOZGTW:** Stop Picking Your Nose, Get To Work
- **STR8:** Straight
- **STRBKS or *$:** Starbucks
- **STS:** So To Speak
- **STW:** Search The Web

- **STYS**: Speak To You Soon
- **SU**: Shut Up
- **SUAC**: Sh** Up A Creek
- **SUAKM**: Shut Up And Kiss Me
- **SUL**: Snooze You Lose
- **SUP**: So What's Up?
- **SUYF**: Shut Up You Fool
- **SWA**: Southwest Airlines
- **SWAG**: Scientific Wild-Ass Guess or Giveaways
- **SWAK**: Sealed With A Kiss
- **SWALBCAKWS**: Sealed With A Lick Because A Kiss Won't Stick
- **SWALK**: Sealed With A Loving Kiss
- **SWDYT**: So What Do You Think?
- **SWF**: Single White Female
- **SWIM**: See What I Mean
- **SWIS**: See What I'm Saying
- **SWL**: Screaming With Laughter
- **SWM**: Single White Male
- **SWMBO**: She Who Must Be Obeyed
- **SWU**: So, What's Up?
- **SYS**: See You Soon
- **SYT**: See You Tomorrow
- **S^**: What's Up?

T

- **T&C**: Terms & Conditions
- **TA**: Thanks Again
- **TABOOMA**: Take A Bite Out Of My Ass
- **TAF**: That's All, Folks!
- **TAFN**: That's All For Now
- **TAH**: Take A Hike
- **TAKS**: That's A Knee-Slapper
- **TAP**: Take A Pill
- **TAS**: Taking A Shower
- **TAW**: Teachers Are Watching
- **TBA**: To Be Advised
- **TBC**: To Be Continued
- **TBD**: To Be Determined
- **TBE**: Thick Between Ears
- **TBH**: To Be Honest
- **TBIL**: The Best Things In Life
- **TBYB**: Try Before You Buy
- **TC**: Take Care
- **TCB**: Trouble Came Back
- **TCOY**: Take Care Of Yourself
- **TD**: The Donald
- **TD&H**: Tall, Dark, & Handsome
- **TDS**: The Dark Side
- **TDTM**: Talk Dirty To Me
- **TEOTWAWKI**: The End of The World As We Know It
- **TFCTR**: The Factor
- **TFDS**: That's For Darn Sure
- **TFH**: Thread From Hell
- **TFN**: Thanks For Nothing or Til Further Notice
- **TFS**: Thanks For Sharing or Three Finger Salute
- **TFTHAOT**: Thanks For The Help Ahead Of Time
- **TFTT**: Thanks For The Thought
- **TGAL**: Think Globally, Act Locally
- **TGGTG**: That Girl/Guy Has Got To Go
- **TGIF**: Thank God It's Friday

- **TGL**: The Good Life
- **THX or TX or THKS**: Thanks
- **TIA**: Thanks In Advance
- **TIAIL**: Think I Am In Love
- **TIC**: Tongue In Cheek
- **TIGAS**: Think I Give A Sh**?
- **TILII**: Tell It Like It Is
- **TINWIS**: That Is Not What I Said
- **TISC**: This Is So Cool
- **TISL**: This Is So Lame
- **TISNC**: This Is So Not Cool
- **TISNF**: That Is So Not Fair
- **TISNT**: That Is So Not True
- **TLC**: Tender Loving Care
- **TLGO**: The List Goes On
- **TLITBC**: That's Life In The Big City
- **TLK2ULTR**: Talk To You Later
- **TM**: Trust Me
- **TMA**: Too Many Acronyms
- **TMI**: Too Much Info
- **TMTOWTDI**: There's More Than One Way To Do It
- **TNA**: Temporarily Not Available
- **TNC**: Tongue In Cheek
- **TNT**: Till Next Time
- **TNTL**: Trying Not To Laugh
- **TNX**: Thanks
- **TOBAL**: There Oughta Be A Law
- **TOBG**: This Oughta Be Good
- **TOF**: Tons Of Fun
- **TOM**: Tomorrow
- **TOPCA**: Till Our Paths Cross Again
- **TOT**: Tons Of Time
- **TOY**: Thinking Of You
- **TP**: Team Player or Teleport
- **TPS**: That's Pretty Stupid
- **TPTB**: The Powers That Be
- **TQM**: Total Quality Management
- **TRAM**: The Rest Are Mine
- **TRDMC**: Tears Running Down My Cheeks
- **TRGT**: Target
- **TRVLRS/TRV**: Travelers Insurance Company
- **TS**: Tough Sh** or Totally Stinks
- **TSIA**: This Says It All
- **TSL**: The Sweet Life
- **TSNF**: That's So Not Fair
- **TSOB**: Tough Son Of A B*tch
- **TSR**: Totally Stupid Rules
- **TSRA**: Two Shakes Of A Rat's Ass
- **TT**: Big Tease
- **TTA**: Tap That Ass
- **TTBOMK**: To The Best Of My Knowledge
- **TTFN**: Ta Ta For Now
- **TTG**: Time To Go
- **TTIOT**: The Truth Is Out There
- **TTKSF**: Trying To Keep A Straight Face
- **TTMF**: Ta Ta, Mofo
- **TTP**: Totally Tragic Perm
- **TTT**: That's The Ticket
- **TTTHTFAI**: Talk To The Hand, The Face Ain't Listening
- **TTTKA**: Time To Totally Kick Ass

- **TTTT:** To Tell The Truth
- **TTUL:** Talk To You Later
- **TTYL:** Talk To You Later
- **TTYT:** Talk To You Tomorrow
- **TWHAB:** This Won't Hurt A Bit
- **TWIMC:** To Whom It May Concern
- **TWTR:** Twitter
- **TWD:** Texting While Driving
- **TXPRT:** Text Expert
- **TXS:** Thanks
- **TXT:** Text Message
- **TXT IM:** Text Instant Message
- **TY:** Thank You
- **TYVM:** Thank You Very Much

U

- **U:** You
- **U Up:** Are You Up?
- **U-L or Ul:** You Will
- **U2:** You Too or U2 (The Band)
- **U8:** You Ate
- **UCWAP:** Up A Creek Without A Paddle
- **UDH82BME:** You'd Hate To Be Me
- **UG2BK:** You've Got To Be Kidding
- **UGC:** User-Generated Content
- **UNPC:** Unpolitically Correct
- **UNTCO:** You Need To Chill Out

- **UOK:** You OK?
- **UPOD:** Underpromise, Overdeliver
- **UR:** You Are
- **UR2K:** You Are Too Kind
- **URAPITA:** You Are A Pain In The Ass
- **URSAI:** You Are Such An Idiot
- **URW:** You Are Welcome
- **URWS:** You Are Wise
- **URYY4M:** You Are Too Wise For Me
- **US:** You Suck
- **USP:** Unique Selling Proposition
- **UTM:** You Tell Me
- **UV:** Unpleasant Visual
- **UWIWU:** You Wish I Was You

V

- **V-AIR:** Virgin Air
- **VBG:** Very Big Grin
- **VBS:** Very Big Smile
- **VC:** Venture Capital
- **VCDA:** Vaya Con Dios, Amigo/Amiga
- **VEG:** Very Evil Grin
- **VFM:** Value For Money
- **VGN:** Vegan or Vegetarian
- **VM:** Voice Mail
- **VRBS:** Virtual-Reality Bullsh**
- **VSF:** Very Sad Face
- **VWD:** Vewry Well Done
- **VWP:** Very Well Played
- **VZN:** Verizon

W

- **W/:** With
- **W/O:** Without
- **W&W:** Wash & Wear
- **W00T:** We Own The Other Team
- **W8:** Wait
- **WAD:** Without A Doubt
- **WAEF:** When All Else Fails
- **WAFM:** What A "F-ing" Mess
- **WAFS:** Warm And Fuzzies
- **WAG:** Wild-Ass Guess
- **WAI:** What An Idiot
- **WAK:** What A Kiss
- **WAYD:** What Are You Doing?
- **WAYN:** Where Are You Now?
- **WB:** Welcome Back or Write Back
- **WBS:** Write Back Soon
- **WC:** Who Cares or Water Closet
- **WCA:** Who Cares Anyway
- **WD:** Well Done
- **WDALYIC:** Who Died And Left You In Charge?
- **WDR:** With All Due Respect
- **WDT:** Who Does That?
- **WDYMBT:** What Do You Mean By That?
- **WDYS:** What Did You Say?
- **WDYT:** What Do You Think?
- **WE:** Whatever
- **WEG:** Wicked Evil Grin
- **WF:** Way Fun
- **WFM:** Works For Me
- **WG:** Wicked Grin

- **WGAFF:** Who Gives A Flying "F"
- **WIBNI:** Wouldn't It Be Nice If
- **WIIFM:** What's In It For Me
- **WILCO:** Will Comply
- **WIM:** Woe Is Me
- **WIP:** Work In Process
- **WISP:** Winning Is So Pleasurable
- **WIT:** Wordsmith In Training
- **WITW:** What In The World
- **WIU:** Wrap It Up
- **WIYW:** What's In Your Wallet?
- **WKEWL:** Way Cool
- **WMHGB:** Where Many Have Gone Before
- **WMMOWS:** Wash My Mouth Out With Soap
- **WMPL:** Wet My Pants Laughing
- **WNOHGB:** Where No One Has Gone Before
- **WNTD:** What Not To Do
- **WNTW:** What Not To Wear
- **WOA:** Work Of Art
- **WOG:** Wise Old Guy
- **WOMBAT:** Waste Of Money, Brains, And Time
- **WOOF:** Well-Off Older Folks
- **WOP:** Without Papers
- **WOTAM:** Waste Of Time And Money
- **WOTD:** Word Of The Day
- **WP:** Well Played
- **WR:** Women Rock or Women Rule

- **WRT:** With Regard To or With Respect To
- **WRUDATM:** What Are You Doing At The Moment?
- **WSID:** What Should I Do?
- **WT:** What The or Who The
- **WTB:** Want To Buy
- **WTF:** What The "F"
- **WTG:** Way To Go
- **WTGP:** Want To Go Private
- **WTH:** What The Heck
- **WTHOW:** White Trash Headline Of The Week
- **WTMI:** Way Too Much Information
- **WTN:** What Then Now
- **WTS:** Want To Sell
- **WTSDS:** Where The Sun Doesnt Shine
- **WTSHTF:** When The Sh** Hits The Fan
- **WTTM:** Without Thinking Too Much
- **WU:** What's Up
- **WUF:** Where You From
- **WUWH:** Wish You Were Here
- **WUWHIMA:** Wish You Were Here In My Arms
- **WWWGD:** What Would A Wise Greek Do?
- **WWWID:** What Would A Wise Italian Do?
- **WWWLD:** What Would A Wise Latino Do?
- **WWJD:** What Would Jesus Do?
- **WWSD:** What Would Satan Do?
- **WWY:** Where Were You?

- **WX:** Weather
- **WYCM?** Will You Call Me?
- **WYGISWYPF:** What You Get Is What You Pay For
- **WYM:** What Do You Mean?
- **WYP:** What's Your Problem?
- **WYRN:** What's Your Real Name?
- **WYS:** Whatever You Say
- **WYSILOB:** What You See Is A Load Of Bull
- **WYSIWYG:** What You See Is What You Get
- **WYSLPG:** What You See Looks Pretty Good
- **WYT:** Whatever You Think
- **WYWH:** Wish You Were Here

- **XI10:** Exciting
- **XLNT:** Excellent
- **XME:** Excuse Me
- **XOXO:** Hugs And Kisses
- **XQZT:** Exquisite
- **XTC:** Ecstasy

- **Y:** Why?
- **YA:** Yet Another
- **YAFIYGI:** You Asked For It, You Got It
- **YAOTM:** Yet Another Off-Topic Message
- **YBF:** You've Been "F-ed"
- **YBS:** You'll Be Sorry
- **YBY:** Yeah Baby Yeah
- **YBYSA:** You Bet Your Sweet Ass

- **YCT**: Your Comment To
- **YDKM**: You Don't Know Me
- **YEPPIES**: Young, Experimenting Perfection-Seekers
- **YGBK**: You Gotta Be Kidding
- **YGBSM**: You Gotta Be Sh**ing Me
- **YGLT**: You're Gonna Love This
- **YGTBK**: You've Got To Be Kidding
- **YGWYPF**: You Get What You Pay For
- **YHM**: You Have Mail
- **YIU**: Yes, I UndersTand
- **YIWGP**: Yes, I Will Go Private
- **YKW**: You Know What
- **YKWIM**: You Know What I Mean
- **YM**: Your Mother, Yo Mama
- **YMAK**: You May Already Know
- **YMMV**: Your Mileage May Vary
- **YNK**: You Never Know
- **YOYO**: You're On Your Own
- **YR**: Yeah Right
- **YS**: You Stinker
- **YSAN**: You're Such A Nerd
- **YSIC**: Why Should I Care?
- **YSK**: You Should Know
- **YSYD**: Yeah, Sure You Do
- **YTB**: You're The Best
- **YTRNW**: Yeah That's Right, Now What?
- **YTTT**: You Telling The Truth?
- **YUPPIES**: Young Urban Professionals
- **YW**: You're Welcome

- **YWIA**: You're Welcome In Advance
- **YYSSW**: Yeah, Yeah, Sure, Sure, Whatever

Z

- **ZZZ**: Sleeping, Bored, Tired

Extras

- **^5**: High Five
- **^RUP^**: Read Up Please
- **^URS**: Up Yours

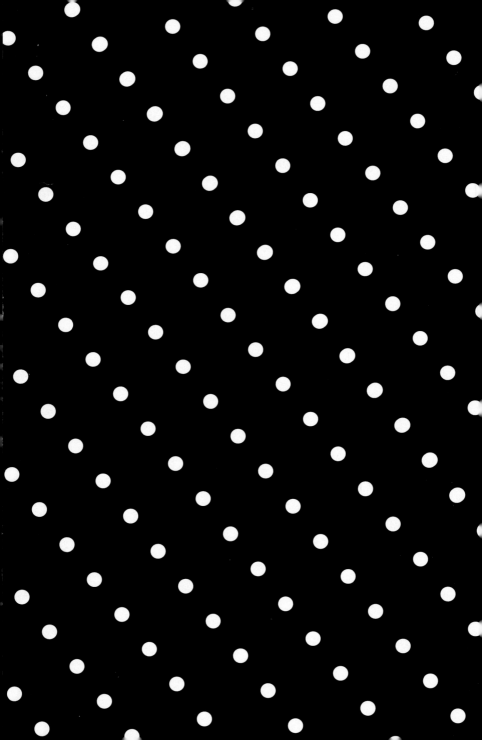

Mind Blowing Texting Info and Fun Facts

Textonomics

Check out these interesting factoids about texting and its exploding growth over the past decade:

- The first SMS text message was sent from the U. K. in December of 1992 by Neil Papworth, an employee of Airwide Solutions and read "Merry Christmas." (www.textually.org)

- In the United States alone, over seventy-five billion text messages are sent/received monthly, with over one trillion sent annually. (CTIA.ORG)

- In the United States, text messaging has exceeded cell phone calls. (Nielson)

- CTIA reports that in the U. S., top line revenues for wireless service providers exceed $150 billion, more than 270 million subscribers.

- According to Wireless Week, three-quarters of text messages are sent between friends, and the remaining quarter are among business contacts.

- Over the next few years, CTIA estimates that over 2.5 trillion text messages will be delivered globally every day.

- According to Nielsen, in 2008, the Obama Campaign's text message announcement was the single largest cell phone marketing event in the U. S., with 2.9 million people receiving texts. If Obama didn't have unlimited texting, that could have cost his campaign $290,000!

- Steve Martin coined NCP over fifteen years ago (nincompoop).

- According to the British Council, regardless of social class, 96 percent of young people own a cell phone in the U. K.

- 47 percent of teenagers say they can accurately text blindfolded. (Nielsen)

- Regis taught Kelly Ripa the fine art of texting—that took some serious skillage! (3-28-2008 broadcast of *Live with Regis and Kelly*)

- What would Bill O'Reilly do? "The Factor" is, you can sign up for the "No Spin" to text your comments.

Now U Can Text

As the years hum by, and as alternative modes of communication become available, you have the freedom to choose which methods best fit your individual lifestyle and needs.

Text messaging fits into my "everything in moderation" category. I've used text messaging in so many different ways, under so many different circumstances, that it's become a time-management staple I can't live without. I use texting daily with my teenagers, mom, friends, coworkers, business partners, and colleagues, all with great success. There have also been a few times where numerous phone call attempts failed me when I was trying to contact a troubled teen. Text messaging helped me successfully establish communications during a panic-filled situation.

The light went on for me when my mom asked me why my kids don't respond to any of her e-mails. Quicker than I could say **OMG**, my response was, "Mom, if you want to communicate with them, you have to text them." After I taught her the text basics, my mom had a response from her granddaughter to her first official text within three seconds! Ever since that day, my mom's been hooked! And since then, she's been successfully texting my teens and everyone else who will text her back. :)

I knew my mom had no idea why my teens wouldn't answer her e-mails, so I felt I had some serious informing to do. With my mom's amazing "you need to write a book and tell everyone about texting" spirit, and a lot of support from my friends and family, I've been able to deliver RBTL.

I've illustrated a few great ways to use texting and when to show restraint. I'm well aware that texting isn't exactly my savior, but I also love the benefits it affords me. It allows me to keep all of my twenty-seven balls in the air at the same time—while staying within etiquette boundaries—without dropping any of them.

Whatever you do, use text messaging with style, grace, and sass . . . and always remember to deliver your best messages. You might even learn to love this amazingly awesome communication tool as much as I have. Go ahead, give it a shot, and shoot for the stars!

The Author
Shawn Marie Edgington

*S*hawn Marie Edgington is a successful CEO and entrepreneur of several multimillion dollar companies. She has spent more than twenty-five years within the insurance industry (www.graniteins.com), and she developed (and nationally launched) E-COMP™ (www.ecomp.us), a unique integrated workers compensation/payroll solution for businesses. Shawn also has more than twenty years of corporate project management, business development, public speaking, brand creation, and creative design and marketing of new and unique services.

Shawn lives in the San Francisco Bay Area, and aside from being a wife and mom, she also makes time to volunteer at school and church, gets her kicks from running, weight training, and kickboxing, and is passionate about healthy living, her two golden retrievers, and traveling. Shawn also stays busy with her two beautiful teenagers, Nicole and Derek, and her husband, David, who is the love of her life and her biggest supporter.

Like most people, Shawn is forever striving to stay balanced with all the demands of life and continues searching for the best time-management tools (yes, texting is one of them) to help her "Shoot for the Stars."

To learn more about RBTL or Shawn, check out www.rbtlguide.com

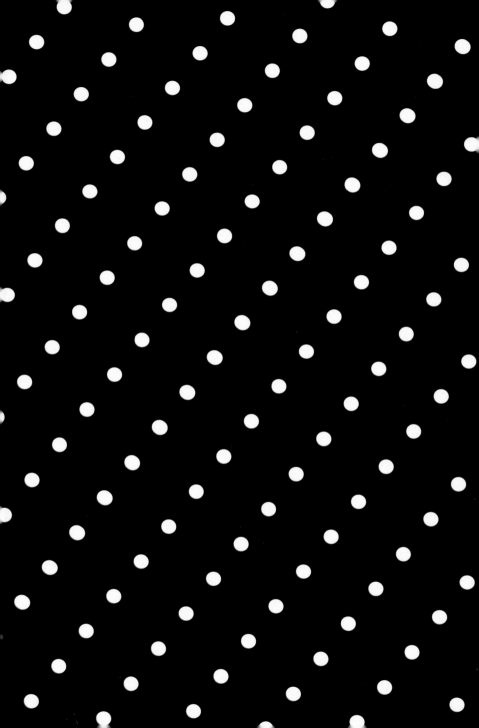

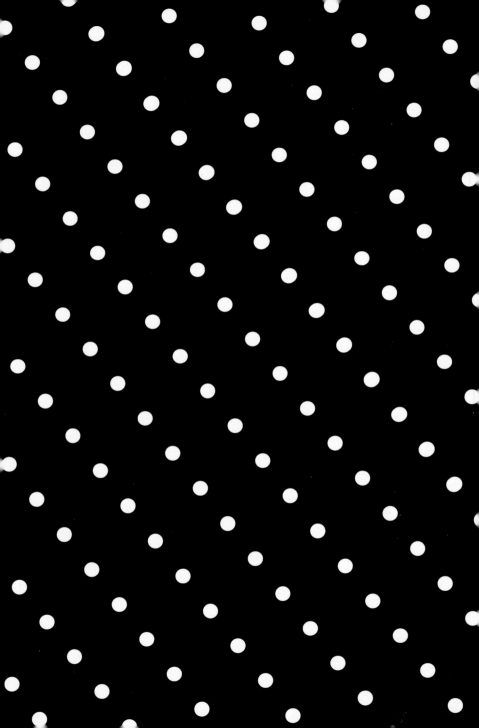